T0212486

SpringerBriefs in Agriculture

More information about this series at http://www.springer.com/series/10183

Vítor João Pereira Domingues Martinho

The Reality for Agricultural Economics Within the European Union

Stressing the Efficiency Indicators Across the Representative Farms

 Springer

Vítor João Pereira Domingues Martinho
Agricultural School
Polytechnic Institute of Viseu
Viseu
Portugal

ISSN 2211-808X   ISSN 2211-8098   (electronic)
SpringerBriefs in Agriculture
ISBN 978-3-319-67010-2   ISBN 978-3-319-67011-9   (eBook)
DOI 10.1007/978-3-319-67011-9

Library of Congress Control Number: 2017951155

Printed on acid-free paper

This Springer imprint is published by Springer Nature
The registered company is Springer International Publishing AG
The registered company address is: Gewerbestrasse 11, 6330 Cham, Switzerland

Contents

The Reality for Agricultural Economics Within the European Union: Stressing the Efficiency Indicators Across the Representative Farms

Abstract Good performance in the agricultural sector is determinant, upon the sensitivity of the sector to the tradeoff against the returns and costs derived from the often small margin of profit in farms. However, this reality diverges between several countries in the European Union, where the reality of the agricultural sector is different, due to various factors, from structural to climatic. Considering that all these countries are subject to common agricultural policies, it is important to understand these contexts through the approach proposed in this study, analyzing microeconomic indicators at farm level and estimating models associated with the convergence theories. In this way, the objective of this study is to analyze several dimensions of the agricultural economics reality for the representative farms of the European Union countries, using microeconomic data available in the Farm Accountancy Data Network, for the period 1989–2009. To support this research, several ratios and efficiency indicators were calculated, across several variables, relatively to, namely, the agricultural output and the utilized agricultural area. These ratios and indicators were analyzed further through econometric models associated with the convergence approaches, presenting the Portuguese specific case. As final remarks, of stressing that it is important to design a set of strategies which are more suited to the European Union realities. In turn, it was verified a tendency of divergence, for the technical efficiency of variable factor, across the twelve former European Union countries, what deserve a special attention from the agricultural operators.

Keywords Farm level · Microeconomic data · Efficiency indicators

Jel Codes C12 · D61 · Q12

A version of this research was considered in the seminar of the aggregation proofs in the Universidade de Trás-os-Montes e Alto Douro, Vila Real, Portugal.

1   Introduction

In the several dimensions of the agricultural economics, stressing the questions related with the technical efficiency of variable factor in the agricultural sector are fundamental, namely because of the reduced profit margin. The importance of this question gains higher dimension in the European Union countries, considering the strategies from the Common Agricultural Policy for all member states. However, these countries have different farming realities, from the structural contexts to the soil and climate frameworks, considering the economic, social and cultural particularities.

In these contexts it is crucial to analyse the agricultural economics realities in the representative European Union farms, not only to support the European and national institutions in the design of adjusted rules, but also to help the agricultural operators, namely the farmers, to choose the better options. There are European Union countries where the farming sector needs special attention, considering specifically the structural problems that bring about serious complications in the trade-off between the returns and the costs (Martinho 2015) and where more adjusted approaches are necessary.

In this framework, the objective of this study is to investigate the dimension of several variables, stressing the questions related to technical efficiency of variable factor, namely taking into account the level of output, in the farms of the European Union member states, using data from the European Union Farm Accountancy Data Network (FADN 2015) for the period from 1989 until 2009. With this statistical information, several ratios and efficiency indicators were calculated and the inter-relationships between variables were analysed, taking into account, namely, the econometric models associated with the convergence theories, presenting the specific case for Portugal.

In terms of technical efficiency of the variable factor, this study is intended as an alternative approach, considering the ratios among the total specific input factors and the total output (from the Cobb-Douglas (1928) production theory, where productivity may be interrelated with technical efficiency) and an analysis of convergence (complemented with pairwise correlation matrix) among the European Union countries and the agrarian regions in the context of Portugal, with data at farm level. In fact, the theory of production considers that the maximum technical efficiency of the variable factor is obtained when the medium productivity is, also, maximum. The calculation of the technical efficiency through the ratio total specific input factors and the total output follows studies such as Martinho (2016) and the Eurostat approach (as presented, for example in: http://ec.europa.eu/eurostat/statistics-explained/index.php/Agri-environmental_indicator_-_energy_use).   This study is justified by the need for more and better research related to the technical efficiency, in European Union countries, as was referred to, for example, by Rezitis and Kalantzi (2016) and for the need of studies as a support toward the design of adjusted policy instruments, as stressed by Giannakis and Bruggeman (2015). There are other approaches to analyse the technical efficiency, as referred to in the following section, such as those related with stochastic frontier analysis, however the methodology presented here aims to be an innovative methodology, using indicators and convergence models, based on microeconomic data from the Farm Accountancy Data Network. In any case, to compare the results, it will be presented the values for the technical efficiency obtained by stochastic frontier analysis of production function.

The European Union Farm Accountancy Data Network provides an interesting database of agricultural microeconomic statistical information for the European Union countries and presents the data for the representative farms through the methodology of weighting described in its website (http://ec.europa.eu/agriculture/rica/). The data available in this database covers the several items related to the various farm dimensions, namely those associated with production, current inputs, investments, subsidies, taxes, assets and cash flow, among others.

The economic theories associated with the absolute convergence and conditional convergences explain both the regional and world evolution, respectively, as a process of economic convergence for the same steady state or different steady states depending on the conditions of each region or country. The absolute (or unconditional) convergence is an economic perspective associated with the neoclassical theory and the conditional convergence, dependent on specific factors of each economy (human qualifications, etc.), appearing to be more related with the endogenous growth theory. It is worth stressing some studies, among others, such as those of Solow (1956), Lucas (1988), Barro (1991), Barro and Sala-i-Martin (1991), Sala-i-Martin (1996) and Romer and Frankel (1999), as references in the convergence theories.

This study is structured in this introduction and in the following sections: the second section for the literature survey (with presentation of the convergence models), the third related to the data analysis through some calculated ratios and the calculation of efficiency indicators, the fourth for econometric results and the main findings conclusions are presented in the last section.

2 Literature Review

The agricultural reality in the European Union countries is, in fact, diverse and with several particularities where the efficient use of several production factors within the farming sector is determinant upon the cost control, helping in the solution of the trade-off to obtain interesting levels of annual profit in the farms.

In these contexts, water is an important farming production factor with great implications upon the growth and productivity of crops (Dorta-Santos et al. 2015). However, the availability of water for farms is not always in abundance, which makes the efficient use of this resource fundamental to farming dynamics (Carvajal et al. 2014).

Another question regarding this resource is pollution control (Wilson 2015), and the consequent effects in the efficiency and in the agricultural economics at farm level. In fact, the impacts of agriculture on water are relevant, in terms of use and degradation of the freshwater available, namely by livestock production, but this can be minimized through more efficient practices, such as changing the feed conversion (Zonderland-Thomassen et al. 2014).

The efficiency in the use of pesticides and other plant protection products is a concern for several researchers and operators, specifically in terms of mitigation of the risks to aquatic environments (Bereswill et al. 2014). A way to reduce the use of pesticides may be the consideration of genetically modified crops (Meissle et al. 2011) which can provide for interesting alternatives, also, in terms of efficiency.

In fact, the problems associated with the use of pesticides in agriculture are critical, in some cases, namely for the water resources, which calls for efficient strategies and instruments of control and mitigation of the inherent consequences for the environment and human health, in preserving the quality of the freshwater (Gregoire et al. 2009).

The efficient utilization of fertilizers is another aspect for water preservation and avoiding problems related to health and sustainability (Lantinga et al. 2013). Several times, it is possible to maintain high levels of economic performance in the farming sector, preserving the sustainability, all of which are dependent upon management tools. Efficiency in the use of fertilizers, specifically those with nitrogen, will also allow for significant reductions in the levels of greenhouse emissions (Rees et al. 2013).

Greenhouse emissions and water pollution are two of the most important direct negative impacts of agriculture on the soil's and air's environment, with indirect consequences across several fields. The main sources of greenhouse emissions are manure management, N_2O soil emissions, cultivation of organic soils, fossil fuel use, fertilizer production and enteric fermentation (Lesschen et al. 2011). The proportion of these sources depends of the farming system, the kind of production and on the efficiency in the use of nutrients.

Energy is another crucial production factor in agriculture, with the appearance of several alternative sources, namely due to the rising prices of fossil fuels and their consequences upon the environment and on sustainability. On the topic of alternative energy sources, in the domain of the named renewable energies, is the woody biomass. This source of energy has been topic of research, specifically in the analysis of the net energy produced (Djomo et al. 2015).

On the other hand, the European agricultural policies and the structural conditions of farms, namely in terms of age and training of the agricultural population, all influence the farming economic performance (Giannakis and Bruggeman 2015), and consequently the efficiency in this economic sector. For example, the structural framework of farms in Finland can bring serious problems of efficiency, with, not only, impacts on costs, but also on the environment (Hiironen and Niukkanen 2014).

Sometimes the efficiency of the strategies implemented in the framework of the European Union Common Agricultural Policy is questioned by authors such as Meyer et al. (2014), namely those related with environmental preservation. In fact, the effects of the European agricultural policies on productivity were negative before the implementation of decoupled subsidies, for the former fifteen European Union countries, and not clear after the decoupling of income financial supports (Rizov et al. 2013). For example, in Slovenia, the persistence of small farms and technical inefficiency is due to the dimension of subsidies driven to the farmers that contribute towards the profitability of Slovenian farms (Bojnec and Latruffe 2013).

Authors, such as Renwick et al. (2013), defend new strategies to avoid land abandonment in the European Union, because the actual existence, of the main pillar or Pillar 1 of the Common Agricultural Policy, is probably technically, economically and environmentally inefficient. Malá (2011), also, concluded that the problems of inefficiency in organic agriculture in the Czech Republic, derived from the implementation of subsidies to support these farming systems.

From another perspective, the efficiency of the policies implemented, in some countries, in order to control the problems caused by nitrogen derived from the use of fertilizers, such as, among others, the Nitrogen-Tax, depends on factors such as the type of farming (Jayet and Petsakos 2013). Similar conclusions were made by Buckley et al. (2012) who concluded that the availability or desire for farmers to implement measures to reduce the environmental impacts from agricultural activity depends on several factors, such as economic and structural dimensions. Farming realities in the European Union are significantly different, which appeals for adjusted strategies for any objective that takes into account the respective particularities, to avoid obtaining unexpected results (Uthes et al. 2010).

For example in Spain, the second greatest reform of the Common Agricultural Policy, verified in 1999, induced a voluntary setting-aside, with consequences in terms of output. At the level of production factors, it favoured the decrease in the use of land, fertilizers and pesticides, and increased the utilization of labour. In general, this reform damaged the technical efficiency in farms (Lambarraa et al. 2009).

In any case, some farming systems and agricultural activities would not be profitable without the subsidies from the Common Agricultural Policy (Amores and Contreras 2009). In the agricultural sector, considering its specificities, efficiency, in some cases such as in familiar agriculture, it is not the most important objective, because this is not a kind of farming with a profitable framework, but with aims more oriented towards social sustainability.

Anyway, the increases in efficiency, with more well-adjusted species and reductions in inputs, may provide interesting ways to find a sustainable economic development and reduce the dependence upon subsidies (Gaspar et al. 2009), considering that perspectives are not promising in these fields of financial support to farms, with the successive enlargements of the European Union. The same opinion has emerged from Sartori et al. (2005) who defend improvements in the farms' efficiency in order to reduce the costs of production and the necessities for public financial support. However, the success in these improvements depends on the kind of agricultural production and on the kind of farming system. Indeed, the impacts of agricultural policies on farm efficiency are different among conventional farms and organic systems. For example, the efficiency in conventional farms is dependent upon the location of the farms in zones of protected quality and has no link to farms with organic productions (Dimara et al. 2005).

The efficiency of the farming policies depends, not only, on the farm and farmers' features, but also, on the perception about the institutions, the considerations of extension services, confidence in the government and the opinion about the stability of the several strategies and instruments in place (Polman and Siangen 2008).

For an effective and efficient application of agricultural policies designed in the European Union, it is important to do adjusted evaluations before their implementation, at several levels of scale, from local to national, avoiding, again, obtaining undesirable outcomes (van Ittersum et al. 2008).

Soil erosion is another question with serious implications upon the efficiency of the agricultural activity (Telles et al. 2011). The soil is one of the most important factors in farming production that when it is not in the best condition reduces the performance of the agricultural activity, with consequences on farmers' returns. Soil conservation appeals for well-designed and well executed adjusted agricultural practices and efficient policies. An effective soil conservation calls for interrelationships between ecological and social contexts (Prager et al. 2011).

The climate can account for part of the changes verified, over the last decades, in the agricultural output. Equally, subsidies, technology and the markets have also made a relevant contribution (Reidsma et al. 2009). However, as already referred to before, the dimension of these interrelationships depends on several factors, such as geographical, cultural, social and management practices.

The effective impacts of climate change in the European agricultural sector will be different across the North and the South, and the West and the East of Europe. The policies must be designed taking into account these questions to mitigate the direct and indirect effects. The multifunctionality in farms can help in these contexts (Olesen and Bindi 2002).

Indeed, a relevant issue for farming efficiency may be the multifunctionality in a context of rural development. In fact, Doucha and Foltýn (2008) identify three aspects for the evaluation of multifunctionality, as being the efficiency, interrelationships with the environment and the interrelationships with rural sustainability.

The harmonization of taxes in the European Union may contribute to improvements in the farming business and make these governmental instruments more efficient, namely value added taxation (David and Nerudová 2008).

Rotational practices are other agricultural options that may improve and optimize efficiency. Indeed, rotations are agricultural needs, because of the conservation of the soils, but it can also be a useful management technique to increase the profitability of the farms (Castellazzi et al. 2007).

The convergence economic theories associated with the absolute and conditional convergence are usually considered in several socioeconomic analyses within different sectors of the economy and by taking into account variables such as the product, productivity and employment, among others. Authors such as Esposti (2011), Hansen and Teuber (2011), Rezitis (2010), McErlean and Wu (2003) and Thirtle et al. (2003) considered the developments from the neoclassical and endogenous growth theories to analyse the convergence tendencies in the agricultural sector in distinct parts of the world.

For example, Esposti (2011) analysed the convergence or divergence of the total factor for productivity in the agricultural sector across Italy, considering data from 1981–1996 and using the three-stage approach, where it was taken into panel unit-root tests and regressions with combined convergence and divergence forces. Hansen and Teuber (2011) studied the convergence of German farmers' revenue and income in the period 1979–2004, combining the methodologies of sigma convergence with Shorrock's inequality decomposition. Rezitis (2010) investigated the agricultural total factor productivity convergence in the United States and in some countries of the European Union from 1973 until 1993, considering the concepts of sigma and beta convergence. McErlean and Wu (2003) examined the agricultural labour productivity convergence in China with conclusions for the period 1985–2000 and, finally, Thirtle et al. (2003) analysed the productivity and the efficiency of agriculture in Botswana for 1981–1996, exploring the two approaches of the concepts of sigma and beta convergence.

The reference to these authors shows that the consideration of productivity as a factor of analysis was preferred for the convergence study in the agricultural sector in different parts of the world. The study presented here for technical efficiency is in line with these studies, considering the relationships between technical efficiency of variable factors and productivity, as stressed in the next section. The convergence approaches will allow us to understand the problems with technical efficiency among the European Union countries, in terms of convergence and divergence, and how agricultural policies can influence this context, considering that one of the assumptions of the European agricultural policies is the socioeconomic convergence between the member states.

2.1 The Convergence Theories Models

Generally with models of convergence theories, related with the neoclassical and endogenous growth approaches, the intention is to analyse the sigma and beta convergence. The sigma convergence measure the dispersion of the variable analysed and the beta convergence is a coefficient of regression across the growth of the variable and its initial value. The beta convergence is a necessary condition, but not sufficient, for the sigma convergence.

Considering the contributions, among others, of Solow (1956), Lucas (1988), Barro (1991), Barro and Sala-i-Martin (1991), Sala-i-Martin (1996), Romer and Frankel (1999), Hansen and Teuber (2011) and Rezitis (2010) the models associated with the convergence theory can be presented as follows:

$$(1/T) \log(Y_{it}/Y_{i0}) = \alpha + (1/T)\left(1 - e^{\beta T}\right)[\log Y_{i0}] + X'\gamma + u_{it}, \quad \text{with}$$
$$\beta < 0 \text{ and } \gamma \neq 0$$

where Y_{it}—variable analysed in region i period t; Y_{i0}—variable analysed in the initial period and region i; T—dimension of the period; α—constant term, representing the steady state; β—coefficient of convergence; u_{it}—error term; X—vector of structural variables for the conditional convergence.

To analyse the convergence with panel data, Islam (1995) proposed an equation obtained considering the Solow (1956) and the Cobb and Douglas (1928) models, with the developments presented below.

$$Y(t) = K(t)^{\alpha}(A(t)L(t))^{1-\alpha} \quad \text{with} \quad 0 < \alpha < 1, \text{ Function of production} \quad (1)$$

$$L(t) = L(0)e^{nt}, \text{ Function of employment} \quad (2)$$

$$A(t) = A(0)e^{gt}, \text{ Function of technical progress} \quad (3)$$

$$\hat{k}(t) = s\hat{y}(t) - (n + g + \delta)\hat{k}(t)$$
$$= s\hat{k}(t)^{\alpha} - (n + g + \delta)\hat{k}(t), \text{ Function of stock of capital} \quad (4)$$

$$\ln y(t_2) - \ln y(t_1) = \gamma \ln y(t_1) + \sum_{j=1}^{2} \beta_j x_{it}^{j} + \eta_t + \mu_i + v_{it} \quad (5)$$

From the Eq. 1, in this model L (labour) and A growth are exogenously obtained with rates n and g, respectively, originating the Eqs. 2 and 3. Considering that s is the constant part of output that is saved and invested and defining the output and the stock of capital per unit of work, such that $\hat{y} = Y/AL$ and $\hat{k} = K/AL$, respectively, the dynamic equation for $\hat{k}$ is given by Eq. 4. From here, taking into account that $\hat{k}$ converges to the value of the steady state, a lambda equal to the elasticities $(\lambda = (n + g + \delta)(1 - \alpha))$ and a set of mathematical transformations it is obtained the Eq. 5 for the convergence analysis with panel data.

3   Data Analysis

Following the literature review made before, it was constructed several tables (Tables A.1–A.20) presented in annex, with values for various ratios and indicators, calculated to better understand the agricultural reality in the twenty seven former European Union countries and the performance (interrelated with the farming reality) of their representative farms. The ratios and indicators were chosen taking into account the statistical information available and with the intention of describe the dynamics and structures of the European Union farms. For the calculation of the efficiency indicators it was used, mostly, the ratio among the variables (namely the specific costs related with variable factors) and the total output, following Martinho (2016) and the developments from the production theory, where the maximum technical efficiency of variable factor is verified when the medium productivity is greater and equal to the marginal productivity.

Data was considered, at farm level, from the FADN (2015), in average for the period 1989–2009 (the longer time series, there are more actualized data until 2013, but only since 2004), divided into two sub-periods for the twelve former countries (1989–1999, for Belgium 1 for example, and 2000–2009, for Belgium 2) and for the three countries that adhered to the European Union in 1995 (1995–1999, for Austria 1 for example, and 2000–2009, for Austria 2). For the countries that adhered after 1995, only one period was considered, from the respective adhesion until 2009. The consideration of the two sub-periods is mainly to capture some changes over the period, namely those associated with the several Common Agricultural Policy reforms.

To improve the analysis of the data, it was stressed in the values, for each variable, the three countries with higher values for the respective ratio/indicator. This methodology it is only to avoid being exhaustive.

3.1 Relationships for the Areas, Livestock Units and Outputs

Concerning the ratios among the area for the several crops and the total utilized agricultural area, of stressing the importance of energy crops, comparatively between the various countries, in Estonia, Lithuania and Slovakia, vegetables and flowers in the Netherlands (two sub-periods), permanent crops and olive groves in Greece, forage crops in Ireland, agricultural fallows in Portugal, agricultural area out of production in Portugal and Spain and set aside in Finland (Table A.1). As referred by Lambarraa et al. (2009) some frameworks, as the Common Agricultural Policy reforms, have impacts in the area used in the agriculture.

In terms of importance of the number of livestock units for the several animal productions relative to the total of livestock units, dairy cows, among the different member-states, are more relevant in Estonia, Latvia and Lithuania, the other cattle in Ireland (second sub-period) and Luxembourg, sheep and goats in Greece, pigs in Belgium (second sub-period) and Denmark and poultry in Hungary, Latvia and Malta (Table A.2). The differences in terms of soil and climate, among the several geographic points in the Europe, have implications in the kind of animal productions adopted.

Relative to the productions of each crop in comparison with the total of crops productions, of highlighting, again, the energy crops in Estonia, Lithuania and Slovakia, vegetables and flowers in the Netherlands and the olive and olive oil in Greece. Of highlighting, also, the relative importance of the oil-seed crops in the Czech Republic, Slovakia and Bulgaria, industrial crops in Greece, citrus fruit in Spain and wine and grapes in Luxembourg (Table A.3).

Animal production for each activity relative to the total livestock production, bearing relevance to the cow' milk and milk products in Finland (first sub-period), Sweden (first sub-period) and Estonia, beef and veal in Ireland, pork in Belgium (second sub-period) and Denmark, sheep and goats in Greece and poultry in Hungary, Malta and Poland (Table A.4).

From these descriptions it is possible to observe the relative importance of the energy crops in Estonia, Lithuania and Slovakia, vegetables and flowers in the Netherlands and the olive productions in Greece. With regard to animal productions, the dairy cow in Estonia, the other cattle productions in Ireland, pork productions in Belgium and Denmark and poultry in Hungary and Malta.

Finally, the farmhouse consumption (comparative to the total output) is higher in Portugal (first sub-period), Slovenia and Romania and the farm use in Estonia, Latvia and Lithuania (Table A.5).

3.2   Ratios Among the Specific and the Common Costs for the Several Agricultural Activities and the Total

The ratio among the total specific costs and the total input costs is higher in Belgium (first sub-period), Malta and Poland. On the other hand, it is worthy stressing the relevance of the specific costs, relative to the total, related with seeds and plants in the Netherlands (second sub-period), Bulgaria and Romania, for fertilizers in Greece (second sub-period), Ireland (second sub-period) and Lithuania, crop protection in France and Greece (first sub-period), feed for grazing livestock in Ireland, Italy and Slovenia, feed for pigs and poultry in Denmark and Malta (Table A.6). Find a balanced point for the tradeoff among the objective of reducing costs and the sustainability will be a big challenge for the farming productions in the future.

The several inputs (relative to the total inputs), are higher for energy in Latvia, Lithuania and Bulgaria, the contract work in Ireland, the interest paid in Denmark, taxes in Italy and VAT on investments in Austria (Table A.7). These different levels of common inputs among the several countries reflect the diverse agricultural economics realities between the member-states.

3.3   Accountancy Variables in Comparison with the Total Utilized Agricultural Area and the Total Output

In this case, of stressing the relative dimension of several accountancy variables in countries as the Netherlands and Malta (Table A.8). This shows the concentration by hectare of assets, for example, in these countries, which may be interesting if incorporating high-end technology.

For the ratios among the accountancy variables and the total output, the context is relatively different (Table A.9). In pointing out, the relative importance of some economic results, such as the gross farm income, farm net value added and farm net income, in Greece (one explanation for this fact can be related with the level of subsidies, what seems plausible, considering the values for the subsidies). Of stressing, the total assets in Ireland, the total current assets in Spain and total liabilities in Denmark. It is relevant to highlight, also, the values for the total and fixed assets in Slovenia. The levels of liabilities in Denmark are in line with the dimension of interest paid by the farms of this country verified before.

3.4 Subsidies Values in Comparison with the Total Utilized Agricultural Area and the Total Output

Relative to the subsidies payments, namely the current subsidies, the values reveal the relative importance (in comparison with the total utilized agricultural area and total output), with some similarities in the two cases, of these values in Austria and Finland (Tables A.10 and A.11). From Oriental and Central European Countries, Malta presents interesting values, namely for the ratio among the subsidy values and the total utilized agricultural area.

On the other hand, the Table A.12 shows the importance relative to the subsidy values for the several animal productions and the total subsidies on livestock. The values show, for the former fifteen European Union countries, differences among the two sub-periods, without primacy for any period.

3.5 Indicators Among Labour, the Total Agricultural Area and the Total Output

It is possible to observe that Greece and Malta are the countries that used more labour per hectare (the relative importance of Greece is mainly for unpaid labour). Cyprus, Malta and Bulgaria use, comparative to the other countries, more paid labour (Tables A.13 and A.14). On the other hand, Portugal and Greece are the countries with more unpaid labour per output and Romania with more unpaid and paid labour.

The values, in comparison with those presented before, for example, for the accountancy variables, show the level of technological development of the representative farms from some European Union countries and the dependence on traditional production factors in other countries.

3.6 Other Ratios and Productivity

The Netherlands and Malta are the countries with a greater performance (Table A.15) for the total output/total utilized agricultural area (the same countries with higher values for the accountancy variables per hectare).

Finland and Slovakia are the two countries with more total input/total output, more total input/total utilized agricultural area in the Netherlands and Malta, more total crop output/total output in Greece, higher livestock output/total output in Ireland, total subsidies on crops/total utilized agricultural area in Greece, total subsidies on livestock/total livestock units in Finland and Malta and total subsidies on crops/crops output and subsidies on livestock/livestock output in Finland. Again, Finland, having significant ratios for subsidies.

On the other hand, from Table A.16 it is possible to observe that the production of cereals/ha and vegetables and flowers/ha are higher in the Netherlands, vineyards in Luxembourg and forage crops in Greece. The traditional European production for wine, such as Spain, France and, by a certain measure, Portugal are not among the leaders of vineyard production by hectare.

Table A.17, about the values for animal production by LU, show that cow´s milk and milk products productions are higher in Finland and Sweden, beef and veal in Italy and the some happens for the pork.

The production of wheat per ha is higher in the Netherlands and the production of milk per cow is greater in Finland and Sweden (Table A.18). Finally, the values confirm that the Netherlands and Malta are the countries with more productions per ha and LU and with more specific costs (Table A.19).

3.7   Efficiency Indicator Obtained with the Ratio Among the Total Specific Costs and the Total Output

Table A.20 shows that in the twelve former European Union countries, in general, the efficiency decrease (due to the ratio increase) from the first to the second period, with the exception of France which remained the same (about 31.5% of the total output for specific costs), the Netherlands which increased (decrease the ratio from 36.4% for 34.4%) and Spain from 32.5% for 31.1%. Greece and Italy are the countries with a higher efficiency, with values around the 22 and 25% mark for the former country and 28 and 29% for the latter member state.

Among the countries that adhered to the European Union in 1995 (Austria, Finland and Sweden), Austria is the country with greater efficiency around 26% and in the other two countries it is close to 45%.

Countries from eastern and central Europe present low efficiencies, in general, with values for the ratio greater than 42%, with the exception of Slovenia with 38.8%. Estonia, Malta and Slovakia are the countries with worse efficiency, presenting values for the ratio at around 50%.

On average the European Union has an efficiency close to 35% which worsened from 34% in the first period to 35% in the second.

Finally, it is worth stressing that after deeper analysis of the information from the FADN (2015), the data shows that there is no relationship between current subsidies and efficiency, for example as in among the twelve former member states.

4 Results

Figure 1 presents the values for the sigma convergence measured through the coefficient of variation, for the technical efficiency of variable factor obtained by the ratio among the total specific costs and the total output (considering this ratio, the better values of efficiency are achieved with low values for the indicator considered). In general there is divergence in the period considered for the twelve former European Union countries (where the data have a longer time series), with signs of some convergence in 1994, 2001 and 2003, curiously these years are close to the more important reforms of the Common Agricultural Policy (first big reform in 1992, second reform in the Agenda 2000 and the intercalary reform of 2003) between 1989 and 2009. In this way, signs of convergence means that the countries with lower ratios (better efficiency) faster growth (losing some efficiency) and those with high ratios (weaker efficiency) slower growth (gaining some efficiency) tending toward an equilibrium. The general tendency of divergence signifies that the countries with great efficiency maintain, compared to those with weak values of technical efficiency.

In Table 1 the results presented are for the beta convergence, using the methodologies available in Stata (2016) for panel data estimations, and taking into account the developments of Islam (1995). Considering the values of the tests used for the cross-sectional dependence, heteroscedasticity and serial autocorrelation, as well as the dimension of the cross-sectional series, the cross-sectional time-series FGLS (feasible generalized least squares) regression was considered. The beta convergence is given by the coefficient among the efficiency growth rate and the logarithm for efficiency in the previous year. The result for the beta convergence

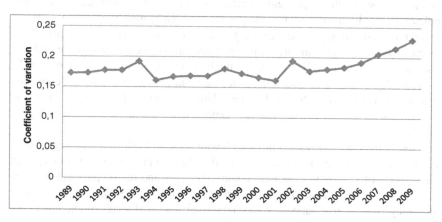

Fig. 1 Sigma convergence among the twelve former European Union countries, for the technical efficiency (total specific costs/total output) over the period 1989–2009

Table 1 Panel data estimation results based on the convergence model (efficiency of variable factor), across the period 1989–2009 and for the twelve former European Union countries

Model	Cross-sectional time-series FGLS regression
Constant	−0.045** (−1.710) [0.087]
Logarithm of efficiency (variable factor) in the previous year	−0.051* (−2.120) [0.034]
Hausman test	42.880* [0.000]
Pesaran's cross-sectional dependence	8.386* [0.000]
Modified Wald test for groupwise	112.480* [0.000]
Wooldridge serial correlation LM test	11.314* [0.006]

Note *Statistically significant at 5%; **Statistically significant at 10%

shows signs of convergence (because it is negative) and around the −0.051 mark. This evidence of beta convergence is not enough to guarantee the sigma convergence.

To confirm the results obtained from panel data other nonparametric estimations were performed, considering the simultaneous quantile regression, for the quantiles 25, 50 and 75. The results, in Table 2, show signs of convergence in the first quantile values of the dependent variable (for the lower growth rate of technical efficiency) and in the second quantile (for the medium growth rate), roughly −0.032 and 0.031, respectively in both cases. This evidence shows that the countries with lower and medium growth rates converge between themselves, a logic of clubs of convergence which is well explained in the literature, for example Chatterji (1992).

Table 3 provides in an ascending order the values, on average, for the efficiency growth rate in the twelve former European Union countries. Italy and Spain are the countries with lower growth rates and Greece and Ireland are those with higher growth.

Cross referencing the values in Table 3 with those in Table A.20 (average values for the efficiency in levels) seems to show that, for example, Italy is a country with good levels of efficiency (low ratio among the costs and the output) and presents low growth rates for this variable, a sign of continuous improvement of technical efficiency. The inverse happens with Greece that has good levels of efficiency, but has the higher growth rates (losing efficiency).

Table 2 Results of absolute convergence for technical efficiency, across the period 1989–2009 and for the twelve former European Union countries, considering quantile regressions

Model	Simultaneous quantile regression
Quantile 25	
Constant	−0.062*
	(−3.270)
	[0.001]
Logarithm of efficiency (variable factor) in the previous year	−0.032*
	(−2.010)
	[0.046]
Quantile 50	
Constant	−0.027
	(−1.560)
	[0.120]
Logarithm of efficiency (variable factor) in the previous year	−0.031*
	(−1.980)
	[0.049]
Quantile 75	
Constant	0.007
	(0.270)
	[0.789]
Logarithm of efficiency (variable factor) in the previous year	−0.032
	(−1.160)
	[0.249]

Note *Statistically significant at 5%

Table 3 Average growth rate of technical efficiency (variable factor), across the period 1989–2009 and for the twelve former European Union countries

Country	Average growth rate of efficiency
Italy	−0.006
Spain	−0.001
Netherlands	0.004
Belgium	0.007
Germany	0.007
France	0.009
Portugal	0.013
United Kingdom	0.013
Denmark	0.014
Luxembourg	0.015
Greece	0.022
Ireland	0.023

4.1　The Sources that May Explain the Differences in the Technical Efficiency of Variable Factor Among the European Union Member States

In order to go deeper and identify the main sources for these differences among efficiency across the European Union countries correlations were analysed between the ratio (total specific costs/total output) considered to quantify the technical efficiency of variable factors and other variables, as can be seen in Tables 4, 5 and 6, through the pairwise correlation matrix and considering all the data available in the FADN (2015) between 1989 and 2009 for the variables studied.

Table 4 shows that manual labour, the area, the number of livestock, fixed assets and the current subsidies have a negative correlation to technical efficiency, considering that the coefficient of correlation among the ratio for total specific costs/total output (the inverse of efficiency) and these variables is positive. The current subsidies are those with a stronger negative correlation to efficiency and the total fixed assets show a lower negative correlation.

On the other hand, Table 5, for the correlation among the technical efficiency and the area (ha) of the several agricultural activities covered, presents that efficiency (inverse of the ratio total specific costs/total output) is positively correlated with the area for vineyards, permanent crops and olive groves (with a coefficient of

Table 4 Pairwise correlation matrix among the technical efficiency of factor variable and other variables, for the twenty seven former European Union countries and for the period 1989–2009

	Technical efficiency	Total labour (h)	Total utilized agricultural área (ha)	Total livestock units (LU)	Total fixed assets (euro)	Total current subsidies (euro)
Technical efficiency	1.000					
Total labour	0.286*	1.000				
	(0.000)					
Total utilized agricultural area	0.386*	0.909*	1.000			
	(0.000)	(0.000)				
Total livestock units	0.379*	0.531*	0.637*	1.000		
	(0.000)	(0.000)	(0.000)			
Total fixed assets	0.245*	0.263*	0.352*	0.762*	1.000	
	(0.000)	(0.000)	(0.000)	(0.000)		
Total current subsidies	0.418*	0.708*	0.855*	0.592*	0.411*	1.000
	(0.000)	(0.000)	(0.000)	(0.000)	(0.000)	

Note *Statistically significant at 5%

Table 5 Pairwise correlation matrix among the technical efficiency and other variables, for the twenty seven former European Union countries and for the period 1989–2009

	Total specific costs/total output	Rented U.A.A. (ha)	Cereals (ha)	Other field crops (ha)	Energy crops (ha)	Vegetables and flowers (ha)	Vineyards (ha)	Permanent crops (ha)	Olive groves (ha)	Orchards (ha)	Other permanent crops (ha)	Forage crops (ha)	Agricultural fallows (ha)	Set aside (ha)	Total agricultural area out of production (ha)	Woodland area (ha)
Total specific costs/total output	1.000															
Rented U.A.A.	0.304* (0.000)	1.000														
Cereals	0.385* (0.000)	0.963* (0.000)	1.000													
Other field crops	0.336* (0.000)	0.982* (0.000)	0.975* (0.000)	1.000												
Energy crops	0.217* (0.000)	0.656* (0.000)	0.625* (0.000)	0.647* (0.000)	1.000											
Vegetables and flowers	0.268* (0.000)	0.473* (0.000)	0.421* (0.000)	0.512* (0.000)	0.253* (0.000)	1.000										
Vineyards	-0.320* (0.000)	0.346* (0.000)	0.266* (0.000)	0.290* (0.000)	0.217* (0.000)	0.030 (0.566)	1.000									
Permanent crops	-0.310* (0.000)	0.095 (0.071)	0.055 (0.292)	0.087 (0.099)	0.107* (0.041)	0.005 (0.925)	0.465* (0.000)	1.000								
Olive groves	-0.438* (0.000)	-0.186* (0.000)	-0.220* (0.000)	-0.208* (0.000)	-0.090 (0.088)	-0.238* (0.000)	0.330* (0.000)	0.924* (0.000)	1.000							
Orchards	-0.053 (0.319)	0.444* (0.000)	0.408* (0.000)	0.455* (0.000)	0.352* (0.000)	0.288* (0.000)	0.555* (0.000)	0.845* (0.000)	0.586* (0.000)	1.000						
Other permanent crops	0.174* (0.001)	0.389* (0.000)	0.388* (0.000)	0.427* (0.000)	0.209* (0.000)	0.582* (0.000)	-0.040 (0.453)	-0.022 (0.670)	-0.215* (0.000)	0.119* (0.023)	1.000					
Forage crops	0.375* (0.000)	0.882* (0.000)	0.844* (0.000)	0.850* (0.000)	0.586* (0.000)	0.433* (0.000)	0.165* (0.002)	-0.088 (0.096)	-0.3070* (0.000)	0.222* (0.000)	0.352* (0.000)	1.000				
Agricultural fallows	0.166* (0.002)	0.514* (0.000)	0.485* (0.000)	0.450* (0.000)	0.137* (0.009)	0.169* (0.001)	0.231* (0.000)	0.219* (0.000)	0.120* (0.021)	0.2846* (0.000)	0.180* (0.001)	0.391* (0.000)	1.000			
Set aside	0.161* (0.002)	0.012 (0.825)	0.146* (0.005)	0.049 (0.355)	-0.032 (0.550)	0.007 (0.899)	-0.051 (0.332)	-0.178* (0.001)	-0.152* (0.004)	-0.171* (0.001)	0.012 (0.820)	0.116* (0.027)	-0.142* (0.007)	1.000		

(continued)

Table 5 (continued)

	Total specific costs/total output	Rented U.A.A. (ha)	Cereals (ha)	Other field crops (ha)	Energy crops (ha)	Vegetables and flowers (ha)	Vineyards (ha)	Permanent crops (ha)	Olive groves (ha)	Orchards (ha)	Other permanent crops (ha)	Forage crops (ha)	Agricultural fallows (ha)	Set aside (ha)	Total agricultural area out of production (ha)	Woodland area (ha)
Total agricultural area out of production	0.242*	0.485*	0.529*	0.445*	0.115*	0.160*	0.188*	0.113*	0.035	0.177*	0.174*	0.426*	0.860*	0.380*	1.000	
	(0.000)	(0.000)	(0.000)	(0.000)	(0.027)	(0.002)	(0.000)	(0.031)	(0.510)	(0.001)	(0.001)	(0.000)	(0.000)	(0.000)		
Woodland area	−0.009	−0.002	−0.002	−0.024	0.071	−0.253*	−0.017	−0.1472*	−0.1409*	−0.091	−0.157*	0.081	0.094	−0.094	0.046	1.000
	(0.868)	(0.971)	(0.970)	(0.646)	(0.176)	(0.000)	(0.745)	(0.005)	(0.007)	(0.084)	(0.003)	(0.125)	(0.073)	(0.075)	(0.381)	

Note: *Statistically significant at 5%

Table 6 Pairwise correlation matrix among the technical efficiency and other variables, for the twenty seven former European Union countries and for the period 1989–2009

	Total specific costs/total output	Dairy cows (LU)	Other cattle (LU)	Sheep and goats (LU)	Pigs (LU)	Poultry (LU)
Total specific costs/total output	1.000					
Dairy cows	0.336*	1.000				
	(0.000)					
Other cattle	0.284*	0.834*	1.000			
	(0.000)	(0.000)				
Sheep and goats	0.151*	0.282*	0.443*	1.000		
	(0.004)	(0.000)	(0.000)			
Pigs	0.324*	0.589*	0.335*	−0.062	1.000	
	(0.000)	(0.000)	(0.000)	(0.238)		
Poultry	0.241*	0.582*	0.332*	0.279*	0.585*	1.000
	(0.000)	(0.000)	(0.000)	(0.000)	(0.000)	

Note *Statistically significant at 5%

correlation among 0.31 for the permanent crops and 0.44 for the olive groves). Maybe this explains the high values of efficiency found for countries such as Italy and Greece. Orchards and woodland area have no correlation with the efficiency ratio and other agricultural production areas are negatively correlated with efficiency (cereals, energy crops, vegetables, flowers, forage crops and set aside, etc.).

Table 6 reveals that the number of livestock units (LU) for the various animal productions are negatively correlated with technical efficiency (inverse of the ratio considered) which is stronger in dairy cows (0.336) and weaker in sheep and goats (0.151).

4.2   Global Technical Efficiency Obtained by Stochastic Frontier Analysis of Production Function

The global technical efficiency obtained by stochastic frontier analysis is the relationship among the potential and the obtained output given the level of outputs (Murteira 1998; Mourão and Martinho 2016). In this analysis it is considered the log Cobb-Douglas production function, where the output at farm level was regressed by the labour, capital (used the total fixed assets) and the productivity (used as proxy the total specific costs), for the former twelve member states (longer time series) and over the period 1989–2009, in panel data.

The results with are those presented in the Table 7. From these values it is possible to verify that the output is mainly explained by the labour and by the total specific costs, showing the importance of the labour in the European Union farming sector. On the other hand the capital has not any relationship with the output. In fact, in the data analysis already was observed that the farming sector in the European Union seems to be more related with the labour, than with the capital.

Table 8 presents the levels of technical efficiency (with values between 0 and 1) for the twelve former member states over the period considered. The values show that, as verified for the technical efficiency of variable factor, countries as Greece and Italy present good performances of technical efficiency.

4.3   The Specific Case of Portugal

Following the methodologies previously presented for the European Union countries, an analysis is presented here of the technical efficiency of variable factor for the Portuguese regions (considering the FADN 2015 data availability), over the period 1989–2007 (for the years 2008 and 2009 the classification of the regions changed).

In the Portuguese regions considered the whole period 1989–2007 was divided into two sub periods, the first from 1989 to 1999 and the second from 2000 to 2007. In this way, for example, Entre Douro e Minho/Beira Litoral1 is for the first sub period (1989–1999) and Entre Douro e Minho/Beira Litoral2 is for the second sub period (2000–2007).

Table 9 shows that the technical efficiency only increased (the ratio decreased) in Trás-os-Montes/Beira Interior (from 0.298 to 0.249). On the other hand, this is the region with greater efficiency (lower ratio). Entre Douro e Minho/Beira Litoral and Ribatejo e Oeste are the Portuguese regions with weaker efficiency (higher

Table 7 Results considering stochastic frontier analysis of production function, across the period 1989–2009 and for the twelve former European Union countries

Model	Stochastic frontier analysis
Constant	1.187* (2.910) [0.004]
Logarithm of total specific costs (proxy for productivity)	0.835* (51.150) [0.000]
Logarithm of labour	0.195* (3.730) [0.000]
Logarithm of total fixed assets (proxy for capital)	0.008 (0.530) [0.595]

Note *Statistically significant at 5%

Table 8 Results for global technical efficiency of output, across the period 1989–2009 and for the twelve former European Union countries, considering stochastic frontier analysis

Country	Technical efficiency
France	0.983996
Greece	0.982389
Netherlands	0.946376
Italy	0.925961
Luxembourg	0.914887
Germany	0.875656
Spain	0.855815
Denmark	0.841578
Belgium	0.779494
United Kingdom	0.757095
Ireland	0.698881
Portugal	0.624192

Table 9 Average values of the technical efficiency (total specific costs/total output) for the Portuguese regions

Region	Total specific costs/total output
Entre Douro e Minho/Beira litoral1	0.394
Entre Douro e Minho/Beira litoral2	0.448
Tras-os-Montes/Beira interior1	0.298
Tras-os-Montes/Beira interior2	0.249
Ribatejo e Oeste1	0.390
Ribatejo e Oeste2	0.394
Alentejo e do Algarve1	0.344
Alentejo e do Algarve2	0.370
Açores e Madeira1	0.339
Açores e Madeira2	0.366

ratio), however, in general, the ratio values are similar between 0.3 and 0.4. In the Entre Douro e Minho/Beira Litoral and in the Ribatejo e Oeste the main productions are mainly dairy milk and horticulture, among others.

In Fig. 2, for the sigma convergence, there are signs of convergence in 1990 and, after 1991 until 1994, after 1998 until 2000, in 2005 and in 2007. As previously referred to for the European Union countries (Fig. 1), it would seem that there is some evidence of convergence around the Common Agricultural Policy reforms.

In general from Fig. 2, there is evidence of divergence for the technical efficiency of variable factor across the Portuguese regions, over the period considered, despite the strong signs of beta convergence presented in Table 10 (−0.207). This, again, shows that beta convergence is not enough to guarantee the sigma convergence in the Portuguese regions.

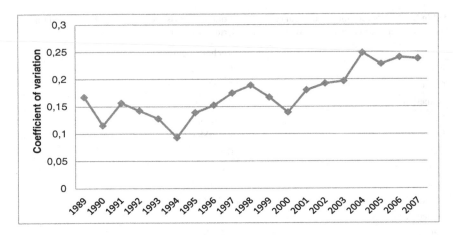

Fig. 2 Sigma convergence among the Portuguese regions, for the technical efficiency, over the period 1989–2007

Table 10 Panel data estimation results based on the convergence model, across the period 1989–2007 and for the Portuguese regions

Model	Cross-sectional time-series FGLS regression
Constant	−0.210* (−2.930) [0.003]
Logarithm of efficiency in the previous year	−0.207* (−3.050) [0.002]
Hausman test	33.750* [0.000]
Pesaran's cross-sectional dependence	1.367 [0.171]
Modified Wald test for groupwise	36.210* [0.000]
Wooldridge serial correlation LM test	268.530* [0.000]

Note *Statistically significant at 5%; **Statistically significant at 10%

5 Conclusions

Considering that the technical efficiency has several dimensions and depends upon a number of variables, the objective of this study was to analyse the agricultural economics reality and the efficiency in the former twenty seven European Union countries, using microeconomic data from the FADN (2015) for the representative farms. With this statistical information several ratios and indicators were calculated to better analyse the data and understand the reality of the representative farms in the European Union countries after decades of Common Agricultural Policy in Europe, stressing the questions related with the technical efficiency (namely in the twelve former European Union countries). To complement this data analysis it was made several estimations based on the convergence models.

The data analysis shows that, comparative to other countries, the energy crops in Estonia, Lithuania and Slovakia, vegetables and flowers in the Netherlands and the olive production in Greece, represent a significant part of the total utilized agricultural area. On the other hand, the dairy cow in Estonia, the other cattle productions in Ireland, pork productions in Belgium and Denmark and poultry in Hungary and Malta, are the animal productions with a greater number of livestock units in total.

The total specific costs, comparative to the total input costs, related with fertilizers are higher in Greece (second sub-period), Ireland (second sub-period) and Lithuania. The efficient utilization of fertilizers is an environmental and sustainability concern stressed, for example, by Lantinga et al. (2013) and Rees et al. (2013). The relative specific costs for crop protection are higher in France and Greece (first sub-period). The issues related with the use of pesticides are another environmental preoccupation as referred to, for example Gregoire et al. (2009), Meissle et al. (2011) and Bereswill et al. (2014). As referred before solve the tradeoff among the efficiency, costs and sustainability it is a big challenge for the European farms in the future.

In terms of accountancy, several variables divided by the total utilized agricultural area presented high values in the Netherlands and Malta (these countries present, also, the greater values for the total output/total utilized agricultural area and total input/total utilized agricultural area). The same variables divided by the total output present different contexts, with some economic results, such as the gross farm income, farm net value added and farm net income, presenting relative importance in Greece (this country also has high values for the total crop output/total output and total subsidies on crops/total utilized agricultural area), the total assets in Ireland (has, also, interesting values for livestock output/total output), the total current assets in Spain and total liabilities in Denmark.

Concerning current subsidy values of stressing their relative importance (comparatively with the total utilized agricultural area and total output), in Austria, Finland and Malta. The questions related to agricultural policies and the efficiency dimensions were analysed, for example, by Meyer et al. (2014), Bojnec and Latruffe (2013) and Rizov et al. (2013).

On the other hand, Finland has, comparatively to other countries, more total input/total output, total subsidies on livestock/total livestock units, total subsidies on crops/crops output, subsidies on livestock/livestock output, cow's milk and milk products production/LU and production of milk per cow. In fact, the questions related with the subsidies assume a relevant dimension in this member-state.

Finally, in the twelve former European Union countries, in general, the efficiency decrease from the years 1989–1999 for the years 2000–2009, exception for France that remain the same, Netherlands and Spain that increase. Greece and Italy are the countries with higher efficiency. Between the countries that adhered to the European Union in 1995, Austria is the country with better efficiency. The countries from the eastern and central Europe presents lower values for the efficiency. Estonia, Malta and Slovakia are the countries with worse efficiency. In the global, the European Union have efficiencies close to 35% that worsened from 34% in the period 1989–1999 to 35% in the period 2000–2009.

The estimations reveal that there are signs of beta convergence in the efficiency among the twelve former European Union countries being around −0.05 and −0.03, depending on the method of regression, which is not enough to guarantee sigma convergence. In fact, evidence was found of divergence from the values of the sigma convergence, which shows that the countries with great technical efficiency of variable factor become better and the others with weak values continue to worsen. The sigma convergence shows that the Common Agricultural Policy reforms seem to have relevant effects upon the convergence temporarily, in the years around these events until one/two years after. On the other hand, the results obtained with the pairwise correlation matrix show that the technical efficiency of variable factor is positively correlated with the area (ha) of vineyards, permanent crops and olive groves, which may explain the levels of efficiency found for Italy and Greece. The results obtained for Portugal follow this tendency, however with greater beta convergence that is still not enough to promote and guarantee signs of sigma convergence. The values for the global technical efficiency, through the stochastic frontier analysis, confirm some tendencies verified for the technical efficiency of variable factor.

In conclusion, there were changes in the two periods considered (1989–1999 and 2000–2009), because it is evident across the data analysis performed in this study, that there are relevant differences between the two periods. This reveals the significant impact of the Common Agricultural Policy reforms upon the European farms, confirmed, namely, by the sigma convergence, that indicated signs of convergence in 1994, 2001 and 2003, respectively years close to the first and second big reforms and the intercalary reform. On the other hand, there are countries where the results seem to be interesting, such as the Netherlands, namely because the levels found for the accountancy variables, normally associated with relevant levels of technological development (as shown at http://www.hollandtradeandinvest.com/key-sectors/agriculture-and-food and https://www.government.nl/topics/agriculture-and-livestock/contents/agriculture-and-horticulture), but are not so good, comparatively with other European Union member states, when we take a deeper look at the technical efficiency ratio. Other countries, such as Austria and Finland, seem to show relevant levels of dependency upon their current subsidies.

In terms of main highlights to be taken into account in the design of future agricultural policies, it is important to stress the need for a set of strategies which are more suited to the European Union realities. For example, some of the European member states with better performance in terms of technical efficiency of variable factor, are those which receives more current subsidies by utilized agricultural area and output. On the other hand, in these contexts, a tendency of divergence was verified, for the technical efficiency of variable factor, across the twelve former

European Union countries. Deeper analysis of the data from the FADN (2015) for these member states concludes that there is not, over the period considered (1989–2009), any relationship between the technical efficiency of variable factor and the level of current subsidies per ha of utilized agricultural area. If one of the concerns of the European Union for the farming sector is to promote the convergence in the technical efficiency, by improving the levels in countries where this indicator is worse, these findings require special attention.

Acknowledgements I would like to thank all those who have contributed in some way to this work.

References

Amores, A. F., & Contreras, I. (2009). New approach for the assignment of new European agricultural subsidies using scores from data envelopment analysis: Application to olive-growing farms in Andalusia (Spain). *European Journal of Operational Research, 193,* 718–729.

Barro, R. (1991). Economic growth in a cross section of countries. *Quarterly Journal of Economics, 106,* 407–501.

Barro, R., & Sala-i-Martin, X. (1991). Convergence across states and regions. *Brooking Papers on Economic Activity, 1,* 82–107.

Bereswill, R., Streloke, M., & Schulz, R. (2014). Risk mitigation measures for diffuse pesticide entry into aquatic ecosystems: Proposal of a guide to identify appropriate measures on a catchment scale. *Integrated Environmental Assessment and Management, 10*(2), 286–298.

Bojnec, S., & Latruffe, L. (2013). Farm size, agricultural subsidies and farm performance in Slovenia. *Land Use Policy, 32,* 207–217.

Buckley, C., Hynes, S., & Mechan, S. (2012). Supply of an ecosystem service—Farmers' willingness to adopt riparian buffer zones in agricultural catchments. *Environmental Science & Policy, 24,* 101–109.

Carvajal, F., Agüera, F., & Sánchez-Hermosilla, J. (2014). Water balance in artificial on-farm agricultural water reservoirs forthe irrigation of intensive greenhouse crops. *Agricultural Water Management, 131,* 146–155.

Castellazzi, M. S., Perry, J. N., Colbach, N., Monod, H., Adamczyk, K., Viaud, V., et al. (2007). New measures and tests of temporal and spatial pattern of crops in agricultural landscapes. *Agriculture, Ecosystems and Environment, 118,* 339–349.

Chatterji, M. (1992). Convergence clubs and Endogenous Growth. *Oxford Review of Economic Policy, 8,* 57–69.

Cobb, C. W., & Douglas, P. H. (1928, March). A theory of production. *The American Economic Review, 18*(1, Supplement). In *Papers and Proceedings of the Fortieth Annual Meeting of the American Economic Association* (pp. 139–165).

David, P., & Nerudová, D. (2008). Selected problems of value added tax application in the agricultural sector of the European Union internal market. *Agricultural Economics—Czech, 54* (1), 1–11.

Dimara, E., Pantzios, C. J., Skuras, D., & Tsekouras, K. (2005). The impacts of regulated notions of quality on farm efficiency: A DEA application. *European Journal of Operational Research, 161,* 416–431.

Djomo, S. N., Ac, A., Zenone, T., Groote, T. D., Bergante, S., Facciotto, G., et al. (2015). Energy performances of intensive and extensive short rotation cropping systems for woody biomass production in the EU. *Renew Sustain Energy Rev, 41*, 845–854.

Dorta-Santos, M., Tejedor, M., Jiménez, C., Hernández-Moreno, J. M., Palacios-Díaz, M. P., & Díaz, F. J. (2015). Evaluating the sustainability of subsurface drip irrigation using recycled wastewater for a bioenergy crop on abandoned arid agricultural land. *Ecological Engineering, 79*, 60–68.

Doucha, T., & Foltýn, I. (2008). Czech agriculture after the accession to the European Union—Impacts on the development of its multifunctionality. *Agricultural Economics—Czech, 54*(4), 150–157.

Esposti, R. (2011). Convergence and divergence in regional agricultural productivity growth: Evidence from Italian regions, 1951–2002. *Agricultural Economics, 42*, 153–169.

FADN. (2015). *Several statistics*. European Commission.

Gaspar, P., Mesías, F. J., Escribano, M., & Pulido, F. (2009). Assessing the technical efficiency of extensive livestock farming systems in Extremadura, Spain. *Livestock Science, 121*, 7–14.

Giannakis, E., & Bruggeman, A. (2015). The highly variable economic performance of European agriculture. *Land Use Policy, 45*, 26–35.

Gregoire, C., Elsaesser, D., Huguenot, D., Lange, J., Lebeau, T., Merli, A., et al. (2009). Mitigation of agricultural nonpoint-source pesticide pollution in artificial wetland ecosystems. *Environmental Chemistry Letters, 7*, 205–231.

Hansen, H., & Teuber, R. (2011). Assessing the impacts of EU's common agricultural policy on regional convergence: Sub-national evidence from Germany. *Applied Economics, 43*, 3755–3765.

Hiironen, J., & Niukkanen, K. (2014). On the structural development of arable land in Finland—How costly will it be for the climate? *Land Use Policy, 36*, 192–198.

Islam, N. (1995). Growth empirics: A panel data approach. *Quarterly Journal of Economics, 110*, 1127–1170.

Jayet, P.-A., & Petsakos, A. (2013). Evaluating the efficiency of a uniform N-Input tax under different policy scenarios at different scales. *Environ Model Assess, 18*, 57–72.

Lambarraa, F., Stefanou, S., Serra, T., & Gil, J. M. (2009). The impact of the 1999 CAP reforms on the efficiency of the COP sector in Spain. *Agricultural Economics, 40*, 355–364.

Lantinga, E. A., Boele, E., & Rabbinge, R. (2013). Maximizing the nitrogen efficiency of a prototype mixed crop-livestock farm in The Netherlands. *NJAS—Wageningen Journal of Life Sciences, 66*, 15–22.

Lesschen, J. P., van den Berg, M., Westhoek, H. J., Witzke, H. P., & Oenema, O. (2011). Greenhouse gas emission profiles of European livestock sectors. *Animal Feed Science and Technology, 166–167*, 16–28.

Lucas, R. E. (1988). On the mechanics of economic development. *Journal of Monetary Economics, 22*, 3–42.

Malá, Z. (2011). Efficiency analysis of Czech organic agriculture. *E+M Ekonomie a Management, 1*, 14–28.

Martinho, V. J. P. D. (Eds.). (2015). *The agricultural economics of the 21st century*. Berlin: Springer.

Martinho, V. J. P. D. (2016). Energy consumption across European Union farms: Efficiency in terms of farming output and utilized agricultural area. *Energy, 103*, 543–556.

McErlean, S., & Wu, Z. (2003). Regional agricultural labour productivity convergence in China. *Food Policy, 28*, 237–252.

Meissle, M., Romeis, J., & Bigler, F. (2011). *Bt* maize and integrated pest management—A European perspective. *Pest Management Science, 67*, 1049–1058.

Meyer, C., Matzdorf, B., Müller, K., & Schleyer, C. (2014). Cross Compliance as payment for public goods? Understanding EU and US agricultural policies. *Ecological Economics, 107*, 185–194.

Mourão, P. R., & Martinho, V. D. (2016). Scoring the efficiency of Portuguese wine exports—An analysis recurring to Stochastic Frontier Models. *Ciência e Técnica Vitivinícola, 31*(1), 1–13.

Murteira, J. (1998). Modelos de fronteira estocástica de produção: Estimação por máxima verosimilhança e avaliação da eficiência técnica. *Estudos de Economia, XVIII, 4*, 491–511.

Olesen, J. E., & Bindi, M. (2002). Consequences of climate change for European agricultural productivity, land use and policy. *European Journal of Agronomy, 16*, 239–262.

Polman, N. B. P., & Siangen, L. H. G. (2008). Institutional design of agri-environmental contracts in the European Union: The role of trust and social capital. *NJAS, 55*(4), 413–430.

Prager, K., Schuler, J., Helming, K., Zander, P., Ratinger, T., & Hagedorn, K. (2011). Soil degradation, farming practices, institutions and policy responses: An analytical framework. *Land Degradation & Development, 22*, 32–46.

Reidsma, P., Lansink, A. O., & Ewert, F. (2009). Economic impacts of climatic variability and subsidies on European agriculture and observed adaptation strategies. *Mitigation and Adaptation Strategies for Global Change, 14*, 35–59.

Rees, R. M., Baddeley, J. A., Bhogal, A., Ball, B. C., Chadwick, D. R., Macleod, M., et al. (2013). Nitrous oxide mitigation in UK agriculture. *Soil Science and Plant Nutrition, 59*(1), 3–15.

Renwick, A., Jansson, T., Verburg, P. H., Revoredo-Giha, C., Britz, W., Gocht, A., et al. (2013). Policy reform and agricultural land abandonment in the EU. *Land Use Policy, 30*, 446–457.

Rezitis, A. N. (2010). Agricultural productivity and convergence: Europe and the United States. *Applied Economics, 42*, 1029–1044.

Rezitis, A. N., & Kalantzi, M. A. (2016). Investigating technical efficiency and its determinants by data envelopment analysis: An application in the greek food and beverages manufacturing industry. *Agribusiness, 32*(2), 254–271.

Rizov, M., Pokrivcak, J., & Ciaian, P. (2013). CAP subsidies and productivity of the EU farms. *Journal of Agricultural Economics, 64*(3), 537–557.

Romer, D., & Frankel, J. A. (1999). Does trade cause growth? *The American Economic Review, 89*(3), 379–399.

Sala-i-Martin, X. (1996). Regional cohesion: Evidence and theories of regional growth and convergence. *European Economic Review, 40*, 1325–1352.

Sartori, L., Basso, B., Bertocco, M., & Oliviero, G. (2005). Energy use and economic evaluation of a three year crop rotation for conservation and organic farming in NE Italy. *Biosystems Engineering, 91*(2), 245–256.

Solow, R. (1956). A contribution to the theory of economic growth. *Quarterly Journal of Economics, 70*(1), 65–94.

Stata. (2016). *Data analysis and statistical software.* http://www.stata.com/.

Telles, T. S., Guimarães, M. F., & Dechen, S. C. F. (2011). The costs of soil erosion. *Revista Brasileira de Ciência do Solo, 35*, 287–298.

Thirtle, C., Piesse, J., Lusigi, A., & Suhariyanto, K. (2003). Multi-factor agricultural productivity, efficiency and convergence in Botswana, 1981–1996. *Journal of Development Economics, 71*, 605–624.

Uthes, S., Sattler, C., Zander, P., Piorr, A., Matzdorf, B., Damgaard, M., et al. (2010). Modeling a farm population to estimate on-farm compliance costs and environmental effects of a grassland extensification scheme at the regional scale. *Agricultural Systems, 103*, 282–293.

van Ittersum, M. K., Ewert, F., Heckelei, T., Wery, J., Olsson, J. A., Andersen, E., et al. (2008). Integrated assessment of agricultural systems—A component-based framework for the European Union (SEAMLESS). *Agricultural Systems, 96*, 150–165.

Wilson, P. (2015). Farm-level actions towards water pollution control: The role of nutrient guidance systems. *Water and Environment Journal, 29*, 88–97.

Zonderland-Thomassen, M. A., Lieffering, M., & Ledgard, S. F. (2014). Water footprint of beef cattle and sheep produced in New Zealand: Water scarcity and eutrophication impacts. *Journal of Cleaner Production, 73*, 253–262.

Website References

http://ec.europa.eu/agriculture/rica/.

http://ec.europa.eu/eurostat/statistics-explained/index.php/Agri-environmental_indicator_-_ener gy_use.

http://www.hollandtradeandinvest.com/key-sectors/agriculture-and-food, https://www.governme nt.nl/topics/agriculture-and-livestock/contents/agriculture-and-horticulture.

Annex

See Tables A.1, A.2, A.3, A.4, A.5, A.6, A.7, A.8, A.9, A.10, A.11, A.12, A.13, A.14, A.15, A.16, A.17, A.18, A.19 and A.20.

© The Author(s) 2017
V.J. Pereira Domingues Martinho, *The Reality For Agricultural Economics Within the European Union*, SpringerBriefs in Agriculture,
DOI 10.1007/978-3-319-67011-9

Table A.1 Ratios among the area occupied by the several crops and the total utilized agricultural area

Country	Rented Utilized Agricultural Area (ha)/Total utilised agricultural area	Cereals (ha)/Total utilised agricultural area	Other field crops (ha)/Total utilised agricultural area	Energy crops (ha)/Total utilised agricultural area	Vegetables and flowers (ha)/Total utilised agricultural area	Vineyards (ha)/Total utilised agricultural area	Permanent crops (ha)/Total utilised agricultural area	Olive groves (ha)/Total utilised agricultural area
Belgium 1	0.756	0.214	0.135	0.000	0.024	0.000	0.012	0.000
Belgium 2	0.745	0.226	0.138	0.003	0.030	0.000	0.017	0.000
Denmark 1	0.227	0.549	0.169	0.000	0.006	0.000	0.004	0.000
Denmark 2	0.269	0.572	0.129	0.006	0.005	0.000	0.004	0.000
France 1	0.756	0.330	0.123	0.000	0.010	0.035	0.008	0.000
France 2	0.832	0.346	0.117	0.008	0.009	0.031	0.008	0.000
Germany 1	0.560	0.392	0.122	0.000	0.005	0.006	0.004	0.000
Germany 2	0.704	0.419	0.150	0.007	0.006	0.005	0.005	0.000
Greece 1	0.299	0.387	0.151	0.000	0.019	0.042	0.277	0.217
Greece 2	0.408	0.384	0.141	0.000	0.017	0.034	0.277	0.228
Ireland 1	0.132	0.064	0.012	0.000	0.001	0.000	0.000	0.000
Ireland 2	0.181	0.062	0.008	0.000	0.000	0.000	0.000	0.000
Italy 1	0.319	0.309	0.066	0.000	0.021	0.061	0.119	0.065
Italy 2	0.385	0.330	0.059	0.000	0.024	0.059	0.114	0.066
Luxembourg 1	0.493	0.226	0.034	0.000	0.000	0.010	0.000	0.000
Luxembourg 2	0.498	0.222	0.052	0.003	0.000	0.010	0.001	0.000
Netherlands 1	0.397	0.089	0.183	0.000	0.050	0.000	0.015	0.000
Netherlands 2	0.401	0.098	0.154	0.000	0.065	0.000	0.017	0.000
Portugal 1	0.324	0.158	0.036	0.000	0.010	0.071	0.109	0.066
Portugal 2	0.318	0.110	0.023	0.000	0.013	0.055	0.108	0.072

(continued)

Table A.1 (continued)

Country	Rented Utilized Agricultural Area (ha)/Total utilised agricultural area	Cereals (ha)/Total utilised agricultural area	Other field crops (ha)/Total utilised agricultural area	Energy crops (ha)/Total utilised agricultural area	Vegetables and flowers (ha)/Total utilised agricultural area	Vineyards (ha)/Total utilised agricultural area	Permanent crops (ha)/Total utilised agricultural area	Olive groves (ha)/Total utilised agricultural area
Spain 1	0.300	0.379	0.090	0.000	0.011	0.037	0.143	0.079
Spain 2	0.333	0.341	0.062	0.000	0.009	0.045	0.144	0.095
United Kingdom 1	0.424	0.226	0.074	0.000	0.006	0.000	0.002	0.000
United Kingdom 2	0.415	0.228	0.081	0.003	0.007	0.000	0.003	0.000
Austria 1	0.265	0.302	0.115	0.000	0.004	0.020	0.009	0.000
Austria 2	0.293	0.308	0.094	0.005	0.004	0.018	0.008	0.000
Finland 1	0.265	0.479	0.078	0.000	0.008	0.000	0.003	0.000
Finland 2	0.341	0.498	0.092	0.001	0.004	0.000	0.004	0.000
Sweden 1	0.463	0.370	0.077	0.000	0.005	0.000	0.001	0.000
Sweden 2	0.498	0.391	0.080	0.003	0.005	0.000	0.000	0.000
Cyprus	0.617	0.404	0.031	0.000	0.038	0.041	0.142	0.069
Czech Republic	0.885	0.460	0.175	0.005	0.006	0.004	0.006	0.000
Estonia	0.608	0.338	0.094	0.011	0.002	0.000	0.003	0.000
Hungary	0.661	0.556	0.180	0.002	0.014	0.011	0.022	0.000
Latvia	0.427	0.341	0.081	0.008	0.007	0.000	0.007	0.000
Lithuania	0.606	0.452	0.117	0.019	0.004	0.000	0.005	0.000
Malta	0.825	0.000	0.102	0.000	0.281	0.101	0.082	0.010

(continued)

Table A.1 (continued)

Country	Rented Utilized Agricultural Area (ha)/Total utilised agricultural area	Cereals (ha)/Total utilised agricultural area	Other field crops (ha)/Total utilised agricultural area	Energy crops (ha)/Total utilised agricultural area	Vegetables and flowers (ha)/Total utilised agricultural area	Vineyards (ha)/Total utilised agricultural area	Permanent crops (ha)/Total utilised agricultural area	Olive groves (ha)/Total utilised agricultural area
Poland	0.281	0.582	0.127	0.001	0.023	0.000	0.023	0.000
Slovakia	0.961	0.382	0.156	0.013	0.003	0.003	0.004	0.000
Slovenia	0.326	0.164	0.032	0.002	0.004	0.025	0.017	0.000
Bulgaria	0.886	0.523	0.244	0.001	0.006	0.025	0.014	0.000
Romania	0.461	0.549	0.161	0.002	0.011	0.008	0.016	0.000
European Union 1	0.466	0.307	0.097	0.000	0.011	0.026	0.052	0.029
European Union 2	0.520	0.350	0.101	0.003	0.011	0.022	0.047	0.028

Country	Orchards (ha)/Total utilised agricultural area	Other permanent crops (ha)/Total utilised agricultural area	Forage crops (ha)/Total utilised agricultural area	Agricultural fallows (ha)/Total utilised agricultural area	Set aside (ha)/Total utilised agricultural area	Total agricultural area out of production (ha)/Total utilised agricultural area	Woodland area (ha)/Total utilised agricultural area
Belgium 1	0.009	0.003	0.597	0.001	0.005	0.006	0.000
Belgium 2	0.015	0.002	0.571	0.001	0.010	0.011	0.000
Denmark 1	0.003	0.001	0.226	0.001	0.045	0.046	0.000
Denmark 2	0.003	0.001	0.227	0.003	0.060	0.063	0.000
France 1	0.008	0.000	0.456	0.007	0.030	0.037	0.015
France 2	0.007	0.000	0.445	0.002	0.039	0.041	0.006

(continued)

Table A.1 (continued)

Country	Orchards (ha)/Total utilised agricultural area	Other permanent crops (ha)/Total utilised agricultural area	Forage crops (ha)/Total utilised agricultural area	Agricultural fallows (ha)/Total utilised agricultural area	Set aside (ha)/Total utilised agricultural area	Total agricultural area out of production (ha)/Total utilised agricultural area	Woodland area (ha)/Total utilised agricultural area
Germany 1	0.003	0.001	0.425	0.007	0.039	0.046	0.081
Germany 2	0.003	0.001	0.367	0.007	0.041	0.048	0.049
Greece 1	0.058	0.003	0.076	0.041	0.002	0.044	0.000
Greece 2	0.048	0.001	0.103	0.032	0.011	0.043	0.002
Ireland 1	0.000	0.000	0.919	0.000	0.004	0.005	0.012
Ireland 2	0.000	0.000	0.922	0.000	0.007	0.007	0.028
Italy 1	0.052	0.002	0.384	0.025	0.011	0.036	0.062
Italy 2	0.042	0.005	0.373	0.020	0.017	0.037	0.081
Luxembourg 1	0.000	0.000	0.723	0.002	0.004	0.006	0.051
Luxembourg 2	0.000	0.000	0.705	0.002	0.008	0.011	0.053
Netherlands 1	0.011	0.004	0.615	0.003	0.006	0.009	0.002
Netherlands 2	0.010	0.008	0.622	0.006	0.010	0.017	0.000
Portugal 1	0.042	0.000	0.302	0.307	0.005	0.312	0.335
Portugal 2	0.037	0.000	0.389	0.290	0.008	0.300	0.199
Spain 1	0.064	0.000	0.222	0.078	0.039	0.117	0.023
Spain 2	0.049	0.000	0.286	0.057	0.054	0.112	0.018
United Kingdom 1	0.002	0.000	0.662	0.006	0.016	0.021	0.016

(continued)

Table A.1 (continued)

Country	Orchards (ha)/Total utilised agricultural area	Other permanent crops (ha)/Total utilised agricultural area	Forage crops (ha)/Total utilised agricultural area	Agricultural fallows (ha)/Total utilised agricultural area	Set aside (ha)/Total utilised agricultural area	Total agricultural area out of production (ha)/Total utilised agricultural area	Woodland area (ha)/Total utilised agricultural area
United Kingdom 2	0.002	0.001	0.629	0.011	0.028	0.039	0.017
Austria 1	0.009	0.000	0.504	0.006	0.039	0.044	0.450
Austria 2	0.008	0.000	0.534	0.006	0.027	0.033	0.372
Finland 1	0.003	0.000	0.346	0.004	0.074	0.078	0.000
Finland 2	0.002	0.001	0.300	0.019	0.074	0.093	0.000
Sweden 1	0.000	0.001	0.474	0.019	0.044	0.063	0.000
Sweden 2	0.000	0.000	0.434	0.011	0.058	0.069	0.000
Cyprus	0.073	0.000	0.239	0.018	0.086	0.104	0.002
Czech Republic	0.006	0.000	0.344	0.003	0.000	0.003	0.002
Estonia	0.003	0.000	0.512	0.022	0.007	0.032	0.152
Hungary	0.020	0.002	0.199	0.006	0.010	0.015	0.039
Latvia	0.006	0.000	0.449	0.095	0.000	0.095	0.158
Lithuania	0.005	0.000	0.354	0.064	0.000	0.066	0.021
Malta	0.071	0.000	0.389	0.043	0.003	0.046	0.000
Poland	0.022	0.001	0.237	0.001	0.007	0.008	0.044
Slovakia	0.003	0.000	0.421	0.028	0.000	0.028	0.002
Slovenia	0.016	0.000	0.754	0.000	0.003	0.003	1.037

(continued)

Table A.1 (continued)

Country	Orchards (ha)/Total utilised agricultural area	Other permanent crops (ha)/Total utilised agricultural area	Forage crops (ha)/Total utilised agricultural area	Agricultural fallows (ha)/Total utilised agricultural area	Set aside (ha)/Total utilised agricultural area	Total agricultural area out of production (ha)/Total utilised agricultural area	Woodland area (ha)/Total utilised agricultural area
Bulgaria	0.014	0.000	0.161	0.022	0.005	0.027	0.000
Romania	0.015	0.000	0.240	0.003	0.010	0.013	0.004
European Union 1	0.023	0.001	0.446	0.033	0.025	0.058	0.045
European Union 2	0.018	0.001	0.408	0.023	0.033	0.056	0.039

Table A.2 Ratios among the number of livestock units for the several animal productions and the total livestock units

Country	Dairy cows (LU)/Total livestock units	Other cattle (LU)/Total livestock units	Sheep and goats (LU)/ Total livestock units	Pigs (LU)/ Total livestock units	Poultry (LU)/Total livestock units
Belgium 1	0.180	0.357	0.001	0.410	0.053
Belgium 2	0.152	0.357	0.002	*0.434*	0.054
Denmark 1	0.166	0.181	0.002	*0.593*	0.054
Denmark 2	0.133	0.143	0.002	*0.639*	0.080
France 1	0.213	0.439	0.053	0.162	0.130
France 2	0.177	0.440	0.051	0.159	0.170
Germany 1	0.277	0.331	0.005	0.363	0.020
Germany 2	0.253	0.293	0.010	0.408	0.034
Greece 1	0.092	0.124	*0.706*	0.034	0.037
Greece 2	0.043	0.107	*0.764*	0.022	0.062
Ireland 1	0.210	0.582	0.131	0.060	0.013
Ireland 2	0.194	*0.646*	0.122	0.021	0.011
Italy 1	0.286	0.341	0.129	0.141	0.087
Italy 2	0.202	0.292	0.092	0.269	0.135
Luxembourg 1	0.309	*0.594*	0.000	0.094	0.001
Luxembourg 2	0.246	*0.622*	0.002	0.114	0.008
Netherlands 1	0.225	0.193	0.017	0.401	0.160
Netherlands 2	0.224	0.182	0.025	0.393	0.172
Portugal 1	0.179	0.339	0.197	0.191	0.068
Portugal 2	0.166	0.387	0.206	0.179	0.044
Spain 1	0.159	0.178	0.266	0.243	0.146
Spain 2	0.091	0.208	0.188	0.374	0.131
United Kingdom 1	0.178	0.383	0.244	0.138	0.056
United Kingdom 2	0.174	0.397	0.226	0.114	0.087
Austria 1	0.272	0.301	0.009	0.372	0.041
Austria 2	0.264	0.317	0.013	0.359	0.041
Finland 1	0.298	0.325	0.010	0.280	0.084
Finland 2	0.277	0.316	0.010	0.295	0.098
Sweden 1	0.278	0.427	0.006	0.277	0.003
Sweden 2	0.242	0.404	0.012	0.316	0.020
Cyprus	0.092	0.069	*0.278*	0.433	0.128
Czech Republic	0.256	0.327	0.005	0.321	0.088
Estonia	*0.335*	0.352	0.034	0.244	0.026
Hungary	0.159	0.173	0.097	0.330	*0.232*

(continued)

Table A.2 (continued)

Country	Dairy cows (LU)/Total livestock units	Other cattle (LU)/Total livestock units	Sheep and goats (LU)/ Total livestock units	Pigs (LU)/ Total livestock units	Poultry (LU)/Total livestock units
Latvia	*0.310*	0.265	0.019	0.213	*0.181*
Lithuania	*0.405*	0.318	0.012	0.152	0.107
Malta	0.160	0.133	0.015	0.418	*0.268*
Poland	0.262	0.183	0.008	0.416	0.114
Slovakia	0.289	0.339	0.061	0.218	0.090
Slovenia	0.245	0.472	0.036	0.178	0.028
Bulgaria	0.244	0.187	0.246	0.139	0.180
Romania	0.290	0.142	0.247	0.143	0.149
European Union 1	0.215	0.348	0.107	0.240	0.085
European Union 2	0.191	0.325	0.095	0.279	0.105

Table A.3 Ratios among the output for the several crops and the total output crop

Country	Cereals (euro)/ Total output crops and crop production	Protein crops (euro)/ Total output crops and crop production	Energy crops (euro)/ Total output crops and crop production	Potatoes (euro)/ Total output crops and crop production	Sugar beet (euro)/ Total output crops and crop production	Oil-seed crops (euro)/ Total output crops and crop production	Industrial crops (euro)/ Total output crops and crop production	Vegetables and flowers (euro)/Total output crops and crop production	Fruit (euro)/ Total output crops and crop production	Citrus fruit (euro)/ Total output crops and crop production	Wine and grapes (euro)/ Total output crops and crop production	Olives and olive oil (euro)/ Total output crops and crop production	Forage crops (euro)/ Total output crops and crop production	Other crop output (euro)/ Total output crops and crop production
Belgium 1	0.167	0.002	0.000	0.084	0.157	0.005	0.013	0.390	0.071	0.000	0.000	0.000	0.001	0.110
Belgium 2	0.135	0.001	0.002	0.094	0.102	0.003	0.024	0.432	0.127	0.000	0.000	0.000	0.008	0.075
Denmark 1	0.499	0.027	0.000	0.053	0.068	0.048	0.001	0.216	0.011	0.000	0.000	0.000	0.009	0.069
Denmark 2	0.449	0.005	0.005	0.056	0.052	0.041	0.001	0.220	0.011	0.000	0.000	0.000	0.077	0.088
France 1	0.346	0.024	0.000	0.019	0.041	0.055	0.008	0.127	0.059	0.000	0.299	0.000	0.005	0.017
France 2	0.303	0.010	0.007	0.029	0.036	0.057	0.011	0.143	0.062	0.001	0.328	0.000	0.000	0.021
Germany 1	0.411	0.004	0.000	0.057	0.120	0.051	0.018	0.159	0.025	0.000	0.090	0.000	0.014	0.050
Germany 2	0.369	0.006	0.006	0.061	0.088	0.089	0.018	0.182	0.032	0.000	0.077	0.000	0.036	0.042
Greece 1	0.146	0.005	0.000	0.018	0.021	0.003	0.199	0.114	0.082	0.045	0.082	0.227	0.043	0.016
Greece 2	0.136	0.012	0.000	0.025	0.013	0.002	0.112	0.161	0.088	0.063	0.082	0.238	0.058	0.010
Ireland 1	0.560	0.004	0.000	0.114	0.152	0.004	0.000	0.038	0.003	0.000	0.000	0.000	0.076	0.048
Ireland 2	0.497	0.006	0.000	0.150	0.082	0.004	0.000	0.029	0.000	0.000	0.000	0.000	0.186	0.046
Italy 1	0.184	0.005	0.000	0.012	0.029	0.014	0.012	0.168	0.109	0.041	0.194	0.082	0.122	0.026
Italy 2	0.145	0.004	0.000	0.014	0.015	0.009	0.005	0.238	0.105	0.028	0.211	0.083	0.095	0.049
Luxembourg 1	0.367	0.007	0.000	0.067	0.000	0.026	0.000	0.000	0.000	0.000	0.517	0.000	0.004	0.012
Luxembourg 2	0.332	0.004	0.005	0.070	0.000	0.058	0.001	0.000	0.002	0.000	0.441	0.000	0.079	0.014
Netherlands 1	0.026	0.002	0.000	0.117	0.049	0.001	0.002	0.694	0.030	0.000	0.000	0.000	0.006	0.075
Netherlands 2	0.019	0.002	0.000	0.082	0.027	0.000	0.005	0.722	0.031	0.000	0.000	0.000	0.007	0.104
Portugal 1	0.181	0.005	0.000	0.074	0.002	0.005	0.008	0.136	0.085	0.019	0.279	0.054	0.094	0.059

(continued)

Table A.3 (continued)

Country	Cereals (euro)/ Total output crops and crop production	Protein crops (euro)/ Total output crops and crop production	Energy crops (euro)/ Total output crops and crop production	Potatoes (euro)/ Total output crops and crop production	Sugar beet (euro)/ Total output crops and crop production	Oil-seed crops (euro)/ Total output crops and crop production	Industrial crops (euro)/ Total output crops and crop production	Vegetables and flowers (euro)/Total output crops and crop production	Fruit (euro)/ Total output crops and crop production	Citrus fruit (euro)/ Total output crops and crop production	Wine and grapes (euro)/ Total output crops and crop production	Olives and olive oil (euro)/ Total output crops and crop production	Forage crops (euro)/ Total output crops and crop production	Other crop output (euro)/ Total output crops and crop production
Portugal 2	0.093	0.002	0.000	0.063	0.010	0.002	0.011	0.215	0.092	0.028	0.282	0.062	0.109	0.031
Spain 1	0.256	0.006	0.000	0.022	0.039	0.027	0.024	0.180	0.109	0.066	0.072	0.133	0.060	0.004
Spain 2	0.201	0.005	0.000	0.024	0.024	0.011	0.014	0.201	0.106	0.080	0.098	0.171	0.050	0.014
United Kingdom 1	0.460	0.022	0.000	0.092	0.082	0.047	0.003	0.179	0.028	0.000	0.000	0.000	0.017	0.071
United Kingdom 2	0.364	0.018	0.004	0.121	0.056	0.055	0.004	0.196	0.034	0.000	0.000	0.000	0.066	0.086
Austria 1	0.363	0.013	0.000	0.056	0.114	0.052	0.019	0.056	0.090	0.000	0.209	0.000	0.012	0.015
Austria 2	0.348	0.007	0.007	0.064	0.092	0.062	0.009	0.057	0.097	0.000	0.225	0.000	0.011	0.028
Finland 1	0.407	0.006	0.000	0.067	0.083	0.021	0.000	0.376	0.009	0.000	0.000	0.000	0.014	0.017
Finland 2	0.385	0.004	0.000	0.090	0.053	0.027	0.003	0.406	0.008	0.000	0.000	0.000	0.013	0.010
Sweden 1	0.579	0.010	0.000	0.134	0.155	0.031	0.007	0.043	0.000	0.000	0.000	0.000	0.028	0.014
Sweden 2	0.444	0.010	0.002	0.099	0.096	0.034	0.001	0.110	0.002	0.000	0.000	0.000	0.188	0.016
Cyprus	0.058	0.005	0.000	0.105	0.000	0.003	0.007	0.382	0.093	0.077	0.067	0.083	0.094	0.027
Czech Republic	0.435	0.005	0.007	0.057	0.066	0.149	0.019	0.062	0.025	0.000	0.026	0.000	0.127	0.031
Estonia	0.414	0.007	0.023	0.063	0.000	0.126	0.000	0.113	0.007	0.000	0.000	0.000	0.248	0.022
Hungary	0.483	0.003	0.001	0.011	0.027	0.142	0.002	0.118	0.055	0.000	0.077	0.000	0.063	0.019
Latvia	0.422	0.003	0.016	0.089	0.033	0.090	0.002	0.117	0.012	0.000	0.000	0.000	0.178	0.054
Lithuania	0.493	0.012	0.028	0.079	0.051	0.104	0.003	0.061	0.017	0.000	0.000	0.000	0.171	0.010
Malta	0.000	0.000	0.000	0.074	0.000	0.000	0.000	0.730	0.036	0.003	0.104	0.002	0.047	0.004
Poland	0.371	0.006	0.001	0.086	0.048	0.069	0.007	0.280	0.085	0.000	0.000	0.000	0.017	0.031
Slovakia	0.485	0.008	0.021	0.031	0.060	0.176	0.004	0.022	0.008	0.000	0.012	0.000	0.152	0.043

(continued)

Table A.3 (continued)

Country	Cereals (euro)/ Total output crops and crop production	Protein crops (euro)/ Total output crops and crop production	Energy crops (euro)/ Total output crops and crop production	Potatoes (euro)/ Total output crops and crop production	Sugar beet (euro)/ Total output crops and crop production	Oil-seed crops (euro)/ Total output crops and crop production	Industrial crops (euro)/ Total output crops and crop production	Vegetables and flowers (euro)/Total output crops and crop production	Fruit (euro)/ Total output crops and crop production	Citrus fruit (euro)/ Total output crops and crop production	Wine and grapes (euro)/ Total output crops and crop production	Olives and olive oil (euro)/ Total output crops and crop production	Forage crops (euro)/ Total output crops and crop production	Other crop output (euro)/ Total output crops and crop production
Slovenia	0.184	0.002	0.001	0.058	0.026	0.008	0.032	0.065	0.087	0.000	0.270	0.004	0.248	0.016
Bulgaria	0.417	0.001	0.001	0.028	0.000	0.174	0.048	0.155	0.044	0.000	0.052	0.000	0.051	0.029
Romania	0.384	0.003	0.001	0.064	0.011	0.087	0.002	0.170	0.074	0.000	0.040	0.000	0.154	0.012
European Union 1	0.277	0.011	0.000	0.042	0.053	0.031	0.023	0.204	0.067	0.019	0.145	0.048	0.043	0.035
European Union 2	0.246	0.007	0.003	0.047	0.039	0.038	0.015	0.238	0.070	0.020	0.140	0.052	0.049	0.041

Table A.4 Ratios among the output for the several animal productions and the total output livestock

Country	Cows' milk and milk products (euro)/Total output livestock and livestock products	Beef and veal (euro)/Total output livestock and livestock products	Pigmeat (euro)/Total output livestock and livestock products	Sheep and goats (euro)/Total output livestock and livestock products	Poultrymeat (euro)/Total output livestock and livestock products	Eggs (euro)/Total output livestock and livestock products	Ewes' and goats' milk (euro)/Total output livestock and livestock products	Other livestock and products (euro)/Total output livestock and livestock products
Belgium 1	0.291	0.282	0.342	0.000	0.061	0.013	0.000	0.010
Belgium 2	0.298	0.243	0.358	0.001	0.073	0.008	0.002	0.018
Denmark 1	0.362	0.104	0.469	0.001	0.029	0.015	0.000	0.020
Denmark 2	0.289	0.059	0.472	0.001	0.041	0.021	0.000	0.117
France 1	0.415	0.285	0.151	0.026	0.060	0.014	0.025	0.024
France 2	0.406	0.270	0.138	0.029	0.068	0.014	0.033	0.041
Germany 1	0.473	0.200	0.288	0.002	0.010	0.012	0.000	0.015
Germany 2	0.504	0.140	0.310	0.003	0.022	0.012	0.000	0.008
Greece 1	0.144	0.095	0.049	0.242	0.041	0.023	0.392	0.015
Greece 2	0.076	0.054	0.020	0.250	0.033	0.036	0.464	0.066
Ireland 1	0.457	0.398	0.060	0.064	0.014	0.002	0.000	0.006
Ireland 2	0.464	0.423	0.021	0.079	0.006	0.001	0.000	0.007
Italy 1	0.492	0.215	0.112	0.040	0.020	0.009	0.068	0.043
Italy 2	0.413	0.179	0.216	0.030	0.016	0.035	0.051	0.060
Luxembourg 1	0.615	0.297	0.083	0.000	0.000	0.003	0.000	0.002
Luxembourg 2	0.592	0.286	0.086	0.000	0.001	0.021	0.003	0.010
Netherlands 1	0.453	0.096	0.299	0.007	0.047	0.075	0.001	0.024
Netherlands 2	0.454	0.060	0.278	0.009	0.054	0.088	0.020	0.035

(continued)

Table A.4 (continued)

Country	Cows' milk and milk products (euro)/Total output livestock and livestock products	Beef and veal (euro)/Total output livestock and livestock products	Pigmeat (euro)/Total output livestock and livestock products	Sheep and goats (euro)/Total output livestock and livestock products	Poultrymeat (euro)/Total output livestock and livestock products	Eggs (euro)/Total output livestock and livestock products	Ewes' and goats' milk (euro)/Total output livestock and livestock products	Other livestock and products (euro)/Total output livestock and livestock products
Portugal 1	0.357	0.230	0.200	0.069	0.057	0.020	0.036	0.030
Portugal 2	0.421	0.199	0.211	0.073	0.022	0.008	0.049	0.017
Spain 1	0.266	0.176	0.202	0.158	0.062	0.023	0.085	0.028
Spain 2	0.217	0.173	0.257	0.113	0.053	0.023	0.109	0.055
United Kingdom 1	0.426	0.240	0.140	0.110	0.021	0.047	0.000	0.016
United Kingdom 2	0.426	0.224	0.098	0.113	0.050	0.052	0.000	0.037
Austria 1	0.403	0.204	0.317	0.004	0.021	0.040	0.000	0.011
Austria 2	0.431	0.193	0.296	0.007	0.028	0.030	0.005	0.010
Finland 1	0.625	0.151	0.161	0.003	0.024	0.035	0.000	0.002
Finland 2	0.613	0.129	0.172	0.003	0.054	0.026	0.000	0.003
Sweden 1	0.642	0.151	0.199	0.002	0.000	0.002	0.000	0.004
Sweden 2	0.578	0.165	0.236	0.005	0.001	0.008	0.000	0.007
Cyprus	0.172	0.027	0.286	0.169	0.077	0.090	0.170	0.010
Czech Republic	0.481	0.143	0.270	0.002	0.063	0.030	0.002	0.009
Estonia	0.620	0.112	0.201	0.012	0.003	0.035	0.001	0.015
Hungary	0.284	0.077	0.253	0.038	0.257	0.054	0.002	0.034

(continued)

Table A.4 (continued)

Country	Cows' milk and milk products (euro)/Total output livestock and livestock products	Beef and veal (euro)/Total output livestock and livestock products	Pigmeat (euro)/Total output livestock and livestock products	Sheep and goats (euro)/Total output livestock and livestock products	Poultrymeat (euro)/Total output livestock and livestock products	Eggs (euro)/Total output livestock and livestock products	Ewes' and goats' milk (euro)/Total output livestock and livestock products	Other livestock and products (euro)/Total output livestock and livestock products
Latvia	0.434	0.089	0.166	0.009	0.048	0.190	0.003	0.059
Lithuania	0.587	0.158	0.146	0.007	0.008	0.088	0.000	0.005
Malta	0.305	0.066	0.260	0.003	0.144	0.165	0.043	0.014
Poland	0.313	0.101	0.317	0.003	0.200	0.037	0.001	0.029
Slovakia	0.558	0.126	0.178	0.011	0.056	0.050	0.015	0.006
Slovenia	0.475	0.284	0.111	0.021	0.001	0.012	0.002	0.096
Bulgaria	0.334	0.085	0.130	0.149	0.099	0.072	0.083	0.049
Romania	0.422	0.070	0.123	0.094	0.075	0.062	0.096	0.059
European Union 1	0.422	0.218	0.212	0.042	0.035	0.024	0.025	0.021
European Union 2	0.402	0.178	0.228	0.041	0.051	0.030	0.034	0.037

Table A.5 Several indicators among other output and the total output

Country	Other output (euro)/ Total output	Farmhouse consumption (euro)/Total output	Farm use (euro)/ Total output
Belgium 1	0.013	0.002	0.022
Belgium 2	0.017	0.001	0.024
Denmark 1	0.032	0.001	0.051
Denmark 2	0.066	0.000	0.058
France 1	0.041	0.006	0.022
France 2	0.045	0.006	0.016
Germany 1	0.105	0.004	0.063
Germany 2	0.094	0.001	0.050
Greece 1	0.004	0.052	0.035
Greece 2	0.019	0.042	0.030
Ireland 1	0.024	0.011	0.032
Ireland 2	0.028	0.004	0.049
Italy 1	0.011	0.017	0.071
Italy 2	0.027	0.009	0.051
Luxembourg 1	0.060	0.006	0.061
Luxembourg 2	0.110	0.003	0.064
Netherlands 1	0.029	0.001	0.009
Netherlands 2	0.072	0.000	0.006
Portugal 1	0.049	*0.061*	0.071
Portugal 2	0.055	0.029	0.064
Spain 1	0.008	0.003	0.035
Spain 2	0.013	0.002	0.024
United Kingdom 1	0.063	0.009	0.022
United Kingdom 2	0.053	0.001	0.042
Austria 1	*0.203*	0.037	0.067
Austria 2	*0.202*	0.025	0.053
Finland 1	0.032	0.009	0.060
Finland 2	0.062	0.004	0.051
Sweden 1	0.082	0.003	0.047
Sweden 2	0.126	0.001	0.102
Cyprus	0.001	0.025	0.040
Czech Republic	0.061	0.002	0.107
Estonia	0.079	0.009	*0.186*
Hungary	0.101	0.003	0.067
Latvia	0.070	0.031	*0.172*
Lithuania	0.017	0.018	*0.168*
Malta	0.008	0.006	0.011

(continued)

Table A.5 (continued)

Country	Other output (euro)/ Total output	Farmhouse consumption (euro)/Total output	Farm use (euro)/ Total output
Poland	0.014	0.017	0.114
Slovakia	*0.165*	0.027	0.137
Slovenia	0.118	*0.063*	0.153
Bulgaria	0.081	0.027	0.061
Romania	0.025	*0.077*	0.099
European Union 1	0.044	0.010	0.040
European Union 2	0.052	0.007	0.041

Table A.6 Ratios among the specific costs for the several productions and the total specific costs

Country	Total specific costs (euro)/ Total Inputs	Seeds and plants (euro)/ Total specific costs	Fertilisers (euro)/ Total specific costs	Crop protection (euro)/ Total specific costs	Other crop specific costs (euro)/Total specific costs	Feed for grazing livestock (euro)/Total specific costs	Feed for pigs and poultry (euro)/Total specific costs	Other livestock specific costs (euro)/ Total specific costs	Forestry specific costs (euro)/ Total specific costs
Belgium 1	0.544	0.078	0.092	0.059	0.043	0.224	0.413	0.091	0.000
Belgium 2	0.507	0.106	0.084	0.070	0.053	0.193	0.386	0.108	0.000
Denmark 1	0.429	0.075	0.091	0.056	0.037	0.205	0.462	0.073	0.001
Denmark 2	0.415	0.067	0.060	0.048	0.039	0.191	0.492	0.099	0.004
France 1	0.377	0.118	0.197	0.162	0.046	0.212	0.199	0.066	0.000
France 2	0.325	0.126	0.202	0.176	0.032	0.215	0.180	0.068	0.000
Germany 1	0.382	0.093	0.119	0.086	0.047	0.268	0.280	0.103	0.002
Germany 2	0.368	0.112	0.127	0.106	0.048	0.239	0.263	0.102	0.001
Greece 1	0.401	0.109	0.211	0.157	0.055	0.382	0.062	0.023	0.000
Greece 2	0.384	0.125	0.232	0.155	0.055	0.352	0.052	0.029	0.000
Ireland 1	0.474	0.030	0.216	0.037	0.026	0.383	0.120	0.188	0.001
Ireland 2	0.447	0.026	0.219	0.038	0.022	0.464	0.040	0.191	0.000
Italy 1	0.472	0.111	0.133	0.097	0.070	0.452	0.104	0.034	0.000
Italy 2	0.438	0.142	0.114	0.098	0.102	0.357	0.142	0.045	0.000
Luxembourg 1	0.384	0.066	0.147	0.053	0.065	0.431	0.101	0.136	0.002
Luxembourg 2	0.349	0.076	0.136	0.078	0.051	0.400	0.107	0.150	0.001
Netherlands 1	0.435	0.148	0.069	0.049	0.118	0.174	0.372	0.069	0.000
Netherlands 2	0.381	0.211	0.045	0.054	0.139	0.138	0.311	0.102	0.000

(continued)

Table A.6 (continued)

Country	Total specific costs (euro)/Total Inputs	Seeds and plants (euro)/Total specific costs	Fertilisers (euro)/Total specific costs	Crop protection (euro)/Total specific costs	Other crop specific costs (euro)/Total specific costs	Feed for grazing livestock (euro)/Total specific costs	Feed for pigs and poultry (euro)/Total specific costs	Other livestock specific costs (euro)/Total specific costs	Forestry specific costs (euro)/Total specific costs
Portugal 1	0.432	0.110	0.138	0.084	0.040	0.379	0.198	0.047	0.003
Portugal 2	0.451	0.100	0.128	0.087	0.070	0.382	0.149	0.078	*0.007*
Spain 1	0.495	0.104	0.179	0.096	0.023	0.346	0.218	0.035	0.000
Spain 2	0.496	0.099	0.162	0.102	0.024	0.328	0.230	0.055	0.000
United Kingdom 1	0.436	0.104	0.146	0.099	0.070	0.281	0.172	0.127	0.000
United Kingdom 2	0.434	0.104	0.139	0.100	0.086	0.304	0.125	0.142	0.000
Austria 1	0.326	0.105	0.103	0.070	0.030	0.187	0.369	0.136	0.000
Austria 2	0.307	0.105	0.096	0.069	0.033	0.244	0.320	0.132	0.002
Finland 1	0.384	0.079	0.191	0.042	0.084	0.306	0.197	0.102	0.000
Finland 2	0.341	0.085	0.157	0.048	*0.112*	0.292	0.200	0.105	0.000
Sweden 1	0.357	0.082	0.149	0.038	0.079	0.434	0.134	0.083	0.000
Sweden 2	0.383	0.077	0.136	0.050	0.059	0.432	0.160	0.085	0.000
Cyprus	0.485	0.091	0.102	0.061	0.014	0.394	0.307	0.030	0.000
Czech Republic	0.407	0.118	0.133	0.132	0.021	0.304	0.215	0.078	0.000
Estonia	0.472	0.101	0.131	0.048	0.055	0.449	0.138	0.075	0.003
Hungary	0.406	0.129	0.144	0.126	0.044	0.194	0.285	0.074	*0.004*

(continued)

Table A.6 (continued)

Country	Total specific costs (euro)/Total Inputs	Seeds and plants (euro)/Total specific costs	Fertilisers (euro)/Total specific costs	Crop protection (euro)/Total specific costs	Other crop specific costs (euro)/Total specific costs	Feed for grazing livestock (euro)/Total specific costs	Feed for pigs and poultry (euro)/Total specific costs	Other livestock specific costs (euro)/Total specific costs	Forestry specific costs (euro)/Total specific costs
Latvia	0.452	0.096	0.139	0.060	0.033	0.346	0.240	0.082	0.004
Lithuania	0.505	0.118	0.240	0.094	0.029	0.356	0.130	0.033	0.000
Malta	0.628	0.079	0.041	0.030	0.039	0.330	0.446	0.035	0.000
Poland	0.520	0.097	0.177	0.085	0.043	0.149	0.408	0.042	0.000
Slovakia	0.361	0.126	0.129	0.133	0.066	0.303	0.136	0.107	0.000
Slovenia	0.380	0.063	0.091	0.052	0.071	0.487	0.131	0.100	0.005
Bulgaria	0.400	0.157	0.188	0.115	0.038	0.286	0.177	0.038	0.001
Romania	0.470	0.159	0.139	0.082	0.048	0.304	0.145	0.124	0.000
European Union 1	0.421	0.106	0.145	0.099	0.057	0.284	0.231	0.077	0.000
European Union 2	0.398	0.119	0.139	0.104	0.061	0.265	0.225	0.085	0.001

Table A.7 Ratios among the several inputs and the total inputs

Country	Total farming overheads (euro)/Total Inputs	Machinery and building current costs (euro)/Total Inputs	Energy (euro)/Total Inputs	Contract work (euro)/Total Inputs	Other direct inputs (euro)/Total Inputs	Depreciation (euro)/Total Inputs	Total external factors (euro)/Total Inputs	Wages paid (euro)/Total Inputs
Belgium 1	0.157	0.046	0.040	0.049	0.021	0.135	0.164	0.035
Belgium 2	0.194	0.053	0.059	0.057	0.025	0.146	0.153	0.049
Denmark 1	0.183	0.083	0.029	0.037	0.033	0.112	0.276	0.076
Denmark 2	0.189	0.068	0.038	0.042	0.041	0.115	*0.281*	0.086
France 1	0.267	0.064	0.044	0.058	0.101	0.170	0.185	0.063
France 2	0.308	0.067	0.047	0.065	*0.128*	0.181	0.186	0.076
Germany 1	0.277	0.101	0.059	0.035	0.081	0.179	0.163	0.066
Germany 2	0.297	0.074	0.078	0.044	0.102	0.143	0.192	0.091
Greece 1	0.223	0.033	0.074	*0.090*	0.025	0.218	0.158	0.067
Greece 2	0.233	0.039	0.091	0.071	0.031	0.221	0.162	0.089
Ireland 1	0.270	*0.117*	0.036	*0.072*	0.045	0.110	0.146	0.046
Ireland 2	0.266	0.109	0.047	*0.075*	0.034	0.166	0.121	0.037
Italy 1	0.187	0.047	0.059	0.038	0.043	0.216	0.125	0.083
Italy 2	0.197	0.037	0.073	0.025	0.062	0.193	0.172	0.129
Luxembourg 1	0.236	0.092	0.034	0.044	0.064	0.255	0.125	0.023
Luxembourg 2	0.245	0.082	0.047	0.047	0.070	0.277	0.129	0.033
Netherlands 1	0.212	0.049	0.062	0.044	0.056	0.153	0.200	0.081
Netherlands 2	0.269	0.057	0.093	0.046	0.073	0.132	0.218	0.104
Portugal 1	0.200	0.051	0.076	0.031	0.043	0.196	0.172	0.125
Portugal 2	0.208	0.054	0.077	0.030	0.046	0.196	0.145	0.111

(continued)

Table A.7 (continued)

Country	Total farming overheads (euro)/Total Inputs	Machinery and building current costs (euro)/Total Inputs	Energy (euro)/Total Inputs	Contract work (euro)/Total Inputs	Other direct inputs (euro)/Total Inputs	Depreciation (euro)/Total Inputs	Total external factors (euro)/Total Inputs	Wages paid (euro)/Total Inputs
Spain 1	0.178	0.037	0.055	0.046	0.040	0.144	0.182	*0.143*
Spain 2	0.215	0.044	0.069	0.042	0.061	0.096	0.192	0.140
United Kingdom 1	0.220	0.075	0.043	0.036	0.065	0.131	0.213	0.123
United Kingdom 2	0.245	0.076	0.055	0.047	0.067	0.119	0.202	0.120
Austria 1	0.292	*0.135*	0.058	0.053	0.046	*0.289*	0.094	0.023
Austria 2	*0.330*	0.109	0.066	0.058	0.097	0.269	0.094	0.025
Finland 1	*0.310*	0.101	0.073	0.018	*0.118*	0.183	0.124	0.053
Finland 2	*0.323*	0.094	0.087	0.040	*0.103*	0.219	0.116	0.049
Sweden 1	0.251	0.092	0.067	0.056	0.036	0.223	0.169	0.050
Sweden 2	0.278	0.079	0.080	0.068	0.050	0.179	0.160	0.058
Cyprus	0.197	0.073	0.076	0.019	0.029	0.171	0.147	0.104
Czech Republic	0.274	0.067	0.090	0.040	0.078	0.089	0.230	*0.190*
Estonia	0.241	0.057	0.101	0.030	0.053	0.116	0.171	0.134
Hungary	0.279	0.040	0.115	0.046	0.078	0.113	0.201	0.129
Latvia	0.277	0.061	*0.130*	0.025	0.061	0.132	0.139	0.100
Lithuania	0.227	0.051	*0.118*	0.009	0.049	0.149	0.120	0.071
Malta	0.207	0.054	0.097	0.028	0.027	0.069	0.096	0.077
Poland	0.223	0.056	0.108	0.030	0.029	0.182	0.074	0.052

(continued)

Table A.7 (continued)

Country	Total farming overheads (euro)/Total Inputs	Machinery and building current costs (euro)/Total Inputs	Energy (euro)/Total Inputs	Contract work (euro)/Total Inputs	Other direct inputs (euro)/Total Inputs	Depreciation (euro)/Total Inputs	Total external factors (euro)/Total Inputs	Wages paid (euro)/Total Inputs
Slovakia	0.251	0.041	0.093	0.047	0.071	0.174	0.215	0.174
Slovenia	0.269	0.132	0.086	0.022	0.028	0.310	0.042	0.022
Bulgaria	0.254	0.043	0.119	0.040	0.052	0.101	0.245	0.138
Romania	0.241	0.051	0.099	0.068	0.024	0.127	0.163	0.112
European Union 1	0.233	0.069	0.052	0.046	0.067	0.167	0.179	0.081
European Union 2	0.260	0.063	0.069	0.048	0.080	0.157	0.184	0.096

Country	Rent paid (euro)/Total Inputs	Interest paid (euro)/Total Inputs	Taxes (euro)/Total Inputs	VAT balance excluding on investments (euro)/Total Inputs	Balance subsidies and taxes on investments (euro)/Total Inputs	Subsidies on investments (euro)/Total Inputs	Payments to dairy outgoers (euro)/Total Inputs	VAT on investments (euro)/Total Inputs
Belgium 1	0.040	0.089	0.007	0.021	−0.015	0.001	0.000	0.016
Belgium 2	0.048	0.057	0.011	0.011	−0.009	0.005	0.000	0.014
Denmark 1	0.026	0.173	0.014	0.000	0.004	0.002	0.002	0.000
Denmark 2	0.038	0.158	0.014	0.000	0.001	0.001	0.000	0.000
France 1	0.069	0.053	0.025	0.000	0.009	0.009	0.000	0.000
France 2	0.076	0.035	0.015	0.000	0.010	0.010	0.000	0.000
Germany 1	0.056	0.040	0.013	0.024	−0.015	0.001	0.001	0.017
Germany 2	0.069	0.032	0.011	0.015	−0.010	0.001	0.000	0.012

(continued)

Table A.7 (continued)

Country	Rent paid (euro)/Total Inputs	Interest paid (euro)/Total Inputs	Taxes (euro)/Total Inputs	VAT balance excluding on investments (euro)/Total Inputs	Balance subsidies and taxes on investments (euro)/Total Inputs	Subsidies on investments (euro)/Total Inputs	Payments to dairy outgoers (euro)/Total Inputs	VAT on investments (euro)/Total Inputs
Greece 1	0.065	0.026	0.003	0.024	0.007	0.010	0.000	0.003
Greece 2	0.069	0.004	0.005	0.001	0.004	0.005	0.000	0.001
Ireland 1	0.045	0.056	0.003	0.000	0.023	0.018	0.004	0.000
Ireland 2	0.054	0.031	0.003	0.004	0.008	0.029	0.000	0.021
Italy 1	0.029	0.013	0.021	0.000	0.008	0.009	0.000	0.001
Italy 2	0.037	0.005	0.026	0.016	0.006	0.007	0.000	0.001
Luxembourg 1	0.050	0.053	0.005	0.052	−0.004	0.038	0.003	0.045
Luxembourg 2	0.051	0.045	0.007	0.044	0.012	0.059	0.000	0.047
Netherlands 1	0.030	0.089	0.007	0.019	−0.009	0.002	0.000	0.012
Netherlands 2	0.035	0.080	0.011	0.002	−0.005	0.001	0.000	0.006
Portugal 1	0.026	0.021	0.003	−0.009	0.053	0.058	0.000	0.006
Portugal 2	0.025	0.009	0.005	−0.009	0.025	0.027	0.001	0.003
Spain 1	0.028	0.011	0.016	0.004	−0.003	0.002	0.000	0.005
Spain 2	0.041	0.010	0.010	0.011	−0.002	0.005	0.000	0.006
United Kingdom 1	0.045	0.045	0.003	0.000	0.004	0.004	0.000	0.000
United Kingdom 2	0.049	0.033	0.005	0.000	0.004	0.003	0.001	0.000
Austria 1	0.032	0.039	0.019	0.036	−0.035	0.026	0.000	0.061
Austria 2	0.036	0.033	0.015	0.050	−0.024	0.025	0.000	0.049
Finland 1	0.025	0.046	0.005	0.000	0.000	0.000	0.000	0.000

(continued)

Table A.7 (continued)

Country	Rent paid (euro)/Total Inputs	Interest paid (euro)/Total Inputs	Taxes (euro)/Total Inputs	VAT balance excluding on investments (euro)/Total Inputs	Balance subsidies and taxes on investments (euro)/Total Inputs	Subsidies on investments (euro)/Total Inputs	Payments to dairy outgoers (euro)/Total Inputs	VAT on investments (euro)/Total Inputs
Finland 2	0.032	0.035	0.002	0.000	0.004	0.004	0.000	0.000
Sweden 1	0.038	0.082	0.001	0.000	0.001	0.000	0.001	0.000
Sweden 2	0.047	0.055	0.001	0.000	0.000	0.000	0.000	0.000
Cyprus	0.032	0.011	0.001	0.000	0.012	0.012	0.000	0.000
Czech Republic	0.028	0.011	0.011	0.000	0.006	0.006	0.000	0.000
Estonia	0.010	0.027	0.004	0.002	0.050	0.051	0.000	0.001
Hungary	0.038	0.033	0.010	-0.001	0.012	0.013	0.000	0.001
Latvia	0.009	0.030	0.007	-0.002	0.050	0.050	0.000	0.000
Lithuania	0.029	0.019	0.006	-0.007	0.116	0.119	0.000	0.002
Malta	0.007	0.012	0.002	0.000	0.004	0.004	0.000	0.000
Poland	0.011	0.012	0.013	0.001	-0.010	0.005	0.000	0.015
Slovakia	0.027	0.013	0.016	0.000	0.013	0.013	0.000	0.000
Slovenia	0.011	0.008	0.002	-0.033	0.033	0.062	0.000	0.029
Bulgaria	0.081	0.026	0.007	0.000	0.004	0.004	0.000	0.000
Romania	0.038	0.012	0.019	-0.002	0.004	0.004	0.000	0.000
European Union 1	0.047	0.051	0.014	0.008	0.001	0.007	0.000	0.006
European Union 2	0.052	0.036	0.012	0.007	0.001	0.007	0.000	0.005

Table A.8 Indicators among accountancy variables and the total utilized agricultural area

Country	Gross Farm Income (euro)/Total utilised agricultural area	Farm Net Value Added (euro)/Total utilised agricultural area	Farm Net Income (euro)/Total utilised agricultural area	Total assets (euro)/Total utilised agricultural area	Total fixed assets (euro)/Total utilised agricultural area	Land, permanent crops and quotas (euro)/Total utilised agricultural area	Buildings (euro)/Total utilised agricultural area	Machinery (euro)/Total utilised agricultural area	Breeding livestock (euro)/Total utilised agricultural area	Total current assets (euro)/Total utilised agricultural area	Non-breeding livestock (euro)/Total utilised agricultural area	Stock of agricultural products (euro)/Total utilised agricultural area
Belgium 1	2165.108	1755.922	1214.801	8961.087	7322.875	2933.460	2042.109	952.727	1394.591	1638.209	1149.224	141.114
Belgium 2	2207.095	1710.016	1158.677	10561.395	8967.269	4332.697	2201.286	1195.478	1237.806	1594.131	964.643	156.696
Denmark 1	1207.023	915.965	206.661	8393.282	6858.416	1672.281	3678.238	1116.992	390.911	1534.864	467.350	215.781
Denmark 2	1395.746	1003.524	18.892	19070.589	16277.855	7839.147	6529.654	1557.764	351.293	2792.732	665.508	342.569
France 1	932.836	693.800	445.766	3998.573	2496.420	892.585	498.227	675.379	430.227	1502.150	272.427	456.886
France 2	994.540	688.311	390.284	4331.433	2640.993	765.927	685.737	773.440	415.888	1690.442	258.245	506.340
Germany 1	1186.780	800.637	420.916	10103.681	8581.660	5243.985	1427.124	1402.657	507.898	1522.018	477.138	56.019
Germany 2	1147.510	832.569	387.137	9351.028	8082.145	5562.952	1111.310	1027.138	380.747	1268.880	354.083	64.527
Greece 1	1940.254	1657.059	1461.967	9668.032	9179.002	7243.698	544.723	1105.224	285.372	489.061	83.788	245.569
Greece 2	2305.160	1932.049	1664.678	10815.750	10305.429	7616.918	723.730	1595.189	369.608	510.276	132.438	124.106
Ireland 1	493.740	416.193	328.543	6447.680	5662.036	4436.079	453.450	292.670	479.842	785.648	468.688	58.521
Ireland 2	644.270	499.758	404.389	15914.536	15063.000	13226.851	869.058	461.468	505.626	851.531	489.363	72.365
Italy 1	1555.707	1242.746	1072.968	17617.354	16229.502	12757.622	1780.751	1329.012	362.099	1387.868	222.023	299.522
Italy 2	2176.347	1781.654	1435.898	21764.530	19737.429	15342.236	2428.182	1615.853	351.149	2027.101	333.138	311.167
Luxembourg 1	1107.196	703.713	498.455	9681.527	8065.384	4295.789	1800.788	1334.206	634.602	1616.141	549.622	190.374
Luxembourg 2	1264.280	736.512	516.130	11893.169	9798.200	5374.329	2031.379	1903.381	489.109	2094.967	597.680	154.747
Netherlands 1	4007.286	2925.965	1449.355	27517.232	23889.007	14114.310	5857.577	2781.044	1136.072	3628.229	1042.134	36.757
Netherlands 2	4525.782	3282.544	1181.758	50156.186	44253.994	33221.062	6424.554	3571.996	1036.378	5902.199	778.932	848.036
Portugal 1	465.519	329.720	247.497	4162.376	3699.922	2387.644	525.508	598.809	187.946	462.469	124.333	123.954
Portugal 2	589.684	438.001	345.631	3490.701	2982.128	1813.382	425.844	539.116	203.797	508.564	108.818	103.723
Spain 1	658.899	564.503	444.677	5770.716	4331.855	3370.148	511.360	241.902	208.442	1438.862	55.354	39.239

(continued)

Table A.8 (continued)

Country	Gross Farm Income (euro)/Total utilised agricultural area	Farm Net Value Added (euro)/Total utilised agricultural area	Farm Net Income (euro)/Total utilised agricultural area	Total assets (euro)/Total utilised agricultural area	Total fixed assets (euro)/Total utilised agricultural area	Land, permanent crops and quotas (euro)/Total utilised agricultural area	Buildings (euro)/Total utilised agricultural area	Machinery (euro)/Total utilised agricultural area	Breeding livestock (euro)/Total utilised agricultural area	Total current assets (euro)/Total utilised agricultural area	Non-breeding livestock (euro)/Total utilised agricultural area	Stock of agricultural products (euro)/Total utilised agricultural area
Spain 2	943.566	864.801	705.182	7649.002	4846.550	3803.755	517.549	295.770	229.487	2802.448	94.235	48.120
United Kingdom 1	606.470	463.342	235.615	5057.193	4357.677	3304.962	228.672	510.881	313.162	699.515	232.898	119.003
United Kingdom 2	668.236	509.463	246.449	7085.944	6179.629	5044.791	240.240	584.726	309.874	906.315	232.634	136.521
Austria 1	1662.438	1206.136	1001.490	11099.847	9922.356	3377.436	4518.947	1740.118	285.876	1177.491	252.715	253.865
Austria 2	1483.688	1037.840	842.475	11280.792	9157.897	2636.970	4522.273	1744.669	253.995	2122.888	245.890	241.443
Finland 1	1049.607	752.188	550.763	5200.501	4107.662	2100.516	944.045	819.606	243.494	1092.839	234.902	195.637
Finland 2	1070.047	665.169	456.595	6337.729	5109.150	2332.935	1387.352	1208.074	180.792	1228.576	165.295	192.104
Sweden 1	526.640	241.979	25.804	4471.679	3739.920	1141.102	1723.402	619.086	256.327	731.759	197.038	96.521
Sweden 2	652.890	377.980	129.380	5860.454	4770.472	2310.114	1475.720	757.625	227.017	1089.982	186.345	152.290
Cyprus	1787.587	1228.749	781.294	22405.536	19414.728	15143.520	1615.289	2015.586	640.380	2990.809	355.327	12.077
Czech Republic	512.396	394.563	102.621	3102.468	2312.584	247.777	1081.970	824.367	158.469	789.884	125.246	134.342
Estonia	284.380	213.628	141.194	1461.627	1144.890	263.701	437.696	349.871	93.620	316.738	63.342	100.066
Hungary	547.390	410.198	181.119	2669.274	1719.904	621.046	480.852	520.493	97.515	949.370	93.754	184.757
Latvia	315.611	231.761	174.905	1394.805	900.863	300.323	201.133	328.344	71.066	493.945	57.334	84.153
Lithuania	371.878	296.952	297.154	1798.774	1208.769	365.309	217.079	542.036	84.352	590.002	57.633	123.096
Malta	8490.802	7510.437	6201.555	84519.028	76450.721	37582.839	25189.143	10455.465	3223.366	8068.316	3599.738	205.785
Poland	734.371	526.231	429.560	5113.012	4331.862	1292.104	1819.717	1059.809	160.243	781.150	159.182	285.527
Slovakia	319.671	142.182	−64.159	2041.697	1494.144	82.346	1120.082	230.682	61.034	547.552	79.731	116.891
Slovenia	1039.443	534.179	515.845	16457.859	15653.968	8942.030	4551.873	1690.508	469.571	803.878	333.379	440.193
Bulgaria	421.345	347.425	170.629	1584.180	1022.247	339.641	251.604	327.139	103.862	561.933	42.896	81.579

(continued)

Table A.8 (continued)

Country	Gross Farm Income (euro)/Total utilised agricultural area	Farm Net Value Added (euro)/Total utilised agricultural area	Farm Net Income (euro)/Total utilised agricultural area	Total assets (euro)/Total utilised agricultural area	Total fixed assets (euro)/Total utilised agricultural area	Land, permanent crops and quotas (euro)/Total utilised agricultural area	Buildings (euro)/Total utilised agricultural area	Machinery (euro)/Total utilised agricultural area	Breeding livestock (euro)/Total utilised agricultural area	Total current assets (euro)/Total utilised agricultural area	Non-breeding livestock (euro)/Total utilised agricultural area	Stock of agricultural products (euro)/Total utilised agricultural area
Romania	626.070	520.975	386.315	3291.832	2588.928	761.697	1137.692	531.348	158.133	702.903	69.925	97.679
European Union 1	1016.574	777.768	523.621	7709.604	6446.087	4287.843	971.787	799.025	387.430	1263.523	279.786	206.344
European Union 2	1102.990	847.259	549.611	8952.164	7364.383	4965.704	1201.163	866.111	331.411	1587.784	250.028	225.003

Country	Other circulating capital (euro)/Total utilised agricultural area	Total liabilities (euro)/Total utilised agricultural area	Long and medium-term loans (euro)/Total utilised agricultural area	Short-term loans (euro)/Total utilised agricultural area	Net worth (euro)/Total utilised agricultural area	Change in net worth (euro)/Total utilised agricultural area	Average farm capital (euro)/Total utilised agricultural area	Gross Investment (euro)/Total utilised agricultural area	Net Investment (euro)/Total utilised agricultural area	Cash Flow (1)(euro)/Total utilised agricultural area	Cash Flow (2)(euro)/Total utilised agricultural area
Belgium 1	347.868	3117.746	3110.151	7.589	5843.341	156.578	6134.439	546.863	137.677	1559.045	1044.670
Belgium 2	472.791	3140.266	3126.216	14.045	7421.126	269.186	6382.609	614.251	117.175	1632.337	957.843
Denmark 1	851.738	4924.139	3622.396	1301.750	3469.142	239.561	6564.819	386.100	95.043	499.893	283.332
Denmark 2	1784.657	10207.577	9168.143	1039.434	8863.009	1035.814	10724.318	858.744	466.519	365.562	488.777
France 1	772.840	1378.923	904.642	474.278	2619.650	74.588	3207.356	247.526	8.484	639.388	440.352
France 2	925.855	1612.009	1027.550	584.458	2719.426	68.924	3669.011	294.016	−12.214	668.098	410.050
Germany 1	988.860	1684.485	1223.744	460.733	8419.199	8.165	4927.488	381.488	−4.653	801.917	434.012
Germany 2	850.273	1642.123	1072.657	569.465	7708.902	31.468	3827.698	342.119	27.179	697.055	388.490
Greece 1	159.733	206.367	146.242	60.125	9461.695	−60.478	4376.436	74.151	−209.105	1601.165	1546.595
Greece 2	253.732	59.661	36.983	22.570	10756.120	−346.829	5306.237	56.371	−316.772	1945.114	1883.362
Ireland 1	258.437	319.276	233.843	85.430	6128.401	167.105	2090.007	107.022	29.475	396.103	306.155
Ireland 2	289.806	420.626	350.416	70.210	15493.910	764.389	2758.131	124.371	−20.138	556.532	451.347
Italy 1	866.306	201.556	179.851	21.722	17415.805	4476.336	5302.995	163.586	−149.382	1325.414	1170.591
Italy 2	1382.773	238.519	217.057	21.470	21526.011	3093.054	7016.064	205.138	−189.582	1784.445	1560.339

(continued)

Table A.8 (continued)

Country	Other circulating capital (euro)/Total utilised agricultural area	Total liabilities (euro)/Total utilised agricultural area	Long and medium-term loans (euro)/Total utilised agricultural area	Short-term loans (euro)/Total utilised agricultural area	Net worth (euro)/Total utilised agricultural area	Change in net worth (euro)/Total utilised agricultural area	Average farm capital (euro)/Total utilised agricultural area	Gross Investment (euro)/Total utilised agricultural area	Net Investment (euro)/Total utilised agricultural area	Cash Flow (1)(euro)/Total utilised agricultural area	Cash Flow (2)(euro)/Total utilised agricultural area
Luxembourg 1	876.148	1291.524	1101.967	189.557	8390.005	181.382	5396.618	564.013	160.528	862.596	366.494
Luxembourg 2	1342.541	1958.759	1689.667	269.092	9934.406	146.731	6513.198	726.188	198.420	1025.859	412.438
Netherlands 1	2549.341	9636.808	8969.368	667.440	17880.428	243.606	13968.380	1545.541	464.213	2475.162	1308.727
Netherlands 2	4275.225	17378.833	14949.428	2429.411	32777.350	543.899	16965.402	1926.768	683.524	2298.953	1380.141
Portugal 1	214.168	124.873	84.205	40.646	4037.502	95.164	2128.511	182.439	46.625	301.042	124.857
Portugal 2	296.039	125.254	67.434	57.829	3365.451	16.152	2048.853	139.384	-12.292	467.433	323.942
Spain 1	1344.262	89.858	71.127	18.722	5680.864	337.050	2941.965	37.391	-56.990	529.473	493.439
Spain 2	2660.087	187.685	164.483	21.614	7461.306	414.579	4005.501	62.617	-16.150	773.744	709.024
United Kingdom 1	347.613	640.085	262.992	377.092	4417.108	84.154	1795.992	159.208	16.079	371.354	228.159
United Kingdom 2	537.159	846.607	400.175	446.433	6239.336	256.247	2030.430	207.806	49.032	395.829	214.582
Austria 1	670.897	1083.526	768.264	315.262	10016.319	30.144	8047.116	554.422	98.113	1314.057	826.518
Austria 2	1635.546	1173.345	851.520	321.828	10107.440	97.700	8796.396	546.492	100.634	1214.046	735.682
Finland 1	662.306	1380.051	1265.058	114.994	3820.450	54.716	3294.916	379.336	81.917	832.431	445.314
Finland 2	871.177	1668.521	1559.741	108.783	4669.205	223.036	4256.165	527.679	122.814	841.171	409.581
Sweden 1	438.200	1427.512	1058.562	368.952	3044.167	-108.327	3476.532	232.222	-52.439	315.942	148.541
Sweden 2	751.344	1853.948	1496.163	357.783	4006.508	152.685	3518.764	334.675	59.764	393.580	134.701
Cyprus	2623.398	501.000	455.198	35.341	21904.536	-353.918	8763.623	131.823	-427.014	1238.646	1061.092
Czech Republic	530.297	713.340	430.093	283.248	2389.128	57.607	2872.943	134.245	16.412	212.949	82.345
Estonia	153.328	408.245	253.706	154.538	1053.382	73.134	1140.697	167.479	96.726	193.033	87.609
Hungary	670.865	746.133	383.479	362.653	1923.141	18.991	2130.951	142.872	5.674	288.349	212.222
Latvia	352.461	420.882	293.389	127.496	973.923	89.514	1029.614	196.286	112.432	224.805	110.179
Lithuania	409.272	276.262	153.773	122.493	1522.512	171.407	1328.466	214.490	139.560	337.032	194.343

(continued)

Table A.8 (continued)

Country	Other circulating capital (euro)/Total utilised agricultural area	Total liabilities (euro)/Total utilised agricultural area	Long and medium-term loans (euro)/Total utilised agricultural area	Short-term loans (euro)/Total utilised agricultural area	Net worth (euro)/Total utilised agricultural area	Change in net worth (euro)/Total utilised agricultural area	Average farm capital (euro)/Total utilised agricultural area	Gross Investment (euro)/Total utilised agricultural area	Net Investment (euro)/Total utilised agricultural area	Cash Flow (1)(euro)/Total utilised agricultural area	Cash Flow (2)(euro)/Total utilised agricultural area
Malta	4262.843	3944.748	2351.953	1593.003	80574.280	418.193	48088.478	1825.370	844.998	7102.321	5543.928
Poland	336.442	483.610	339.202	144.380	4629.392	149.757	3900.338	222.213	14.082	593.535	382.867
Slovakia	350.930	173.116	88.606	84.509	1868.580	−74.335	2026.035	120.730	−56.759	97.852	−5.539
Slovenia	30.275	312.042	303.051	8.992	16145.817	204.760	7797.341	648.601	143.309	862.192	231.854
Bulgaria	437.469	303.194	197.734	105.447	1281.010	144.261	1294.158	161.305	87.398	209.187	84.981
Romania	535.358	102.241	61.088	41.181	3189.591	266.189	2495.098	54.571	−50.523	374.901	327.310
European Union 1	777.399	1093.637	794.613	299.028	6615.969	577.930	3690.608	229.138	−9.658	729.045	532.395
European Union 2	1112.759	1320.105	978.718	341.140	7632.062	474.969	4159.475	260.286	4.557	778.521	569.893

Table A.9 Indicators among accountancy variables and the total output

Country	Gross Farm Income (euro)/Total output	Farm Net Value Added (euro)/Total output	Farm Net Income (euro)/Total output	Total assets (euro)/Total output	Total fixed assets (euro)/Total output	Land, permanent crops and quotas (euro)/Total output	Buildings (euro)/Total output	Machinery (euro)/Total output	Breeding livestock (euro)/Total output	Total current assets (euro)/Total output	Non-breeding livestock (euro)/Total output	Stock of agricultural products (euro)/Total output
Belgium 1	0.538	0.436	0.301	2.232	1.825	0.731	0.509	0.238	0.347	0.407	0.286	0.035
Belgium 2	0.530	0.411	0.278	2.525	2.142	1.031	0.527	0.286	0.298	0.382	0.232	0.037
Denmark 1	0.457	0.347	0.077	3.198	2.616	0.639	1.404	0.424	0.148	0.583	0.177	0.082
Denmark 2	0.444	0.319	0.012	5.896	5.027	2.312	2.113	0.490	0.111	0.869	0.210	0.105
France 1	0.556	0.414	0.266	2.380	1.487	0.531	0.297	0.402	0.257	0.893	0.162	0.272
France 2	0.570	0.394	0.223	2.481	1.513	0.439	0.393	0.443	0.238	0.968	0.148	0.290
Germany 1	0.504	0.342	0.179	4.331	3.694	2.296	0.600	0.585	0.214	0.637	0.200	0.024
Germany 2	0.515	0.373	0.173	4.216	3.645	2.510	0.501	0.462	0.172	0.571	0.160	0.029
Greece 1	0.820	0.700	0.618	4.089	3.882	3.065	0.230	0.466	0.121	0.207	0.035	0.104
Greece 2	0.877	0.735	0.633	4.127	3.933	2.907	0.276	0.609	0.141	0.194	0.050	0.047
Ireland 1	0.567	0.478	0.376	7.445	6.541	5.132	0.522	0.337	0.551	0.903	0.539	0.067
Ireland 2	0.729	0.566	0.459	17.937	16.972	14.898	0.979	0.520	0.574	0.965	0.555	0.082
Italy 1	0.649	0.519	0.448	7.372	6.793	5.342	0.745	0.555	0.151	0.578	0.093	0.125
Italy 2	0.693	0.567	0.458	7.105	6.455	5.043	0.778	0.521	0.113	0.650	0.107	0.100
Luxembourg 1	0.606	0.384	0.270	5.332	4.445	2.380	0.988	0.733	0.345	0.887	0.303	0.104
Luxembourg 2	0.689	0.401	0.281	6.483	5.340	2.931	1.106	1.037	0.267	1.142	0.326	0.085
Netherlands 1	0.476	0.347	0.172	3.272	2.841	1.679	0.697	0.330	0.135	0.431	0.124	0.004
Netherlands 2	0.439	0.320	0.118	4.871	4.307	3.246	0.621	0.339	0.101	0.565	0.076	0.078
Portugal 1	0.566	0.400	0.298	5.110	4.543	2.930	0.645	0.736	0.231	0.567	0.153	0.152
Portugal 2	0.649	0.482	0.380	3.849	3.291	2.004	0.470	0.593	0.224	0.559	0.120	0.114
Spain 1	0.669	0.574	0.453	5.831	4.372	3.403	0.516	0.243	0.210	1.459	0.055	0.040
Spain 2	0.719	0.659	0.537	5.813	3.688	2.892	0.395	0.226	0.175	2.125	0.072	0.037

(continued)

Table A.9 (continued)

Country	Gross Farm Income (euro)/Total output	Farm Net Value Added (euro)/Total output	Farm Net Income (euro)/Total output	Total assets (euro)/Total output	Total fixed assets (euro)/Total output	Land, permanent crops and quotas (euro)/Total output	Buildings (euro)/Total output	Machinery (euro)/Total output	Breeding livestock (euro)/Total output	Total current assets (euro)/Total output	Non-breeding livestock (euro)/Total output	Stock of agricultural products (euro)/Total output
United Kingdom 1	0.519	0.397	0.203	4.316	3.718	2.820	0.195	0.436	0.268	0.598	0.199	0.102
United Kingdom 2	0.514	0.391	0.188	5.439	4.742	3.869	0.184	0.451	0.239	0.697	0.179	0.105
Austria 1	0.836	0.606	0.503	5.582	4.989	1.682	2.284	0.879	0.144	0.593	0.127	0.128
Austria 2	0.778	0.543	0.441	5.903	4.802	1.377	2.376	0.915	0.134	1.101	0.127	0.127
Finland 1	0.767	0.550	0.404	3.802	3.004	1.540	0.686	0.597	0.180	0.798	0.173	0.143
Finland 2	0.756	0.470	0.324	4.463	3.596	1.640	0.976	0.851	0.128	0.868	0.117	0.134
Sweden 1	0.484	0.222	0.024	4.088	3.421	1.049	1.569	0.569	0.233	0.667	0.178	0.087
Sweden 2	0.484	0.275	0.091	4.347	3.554	1.699	1.121	0.565	0.169	0.793	0.139	0.111
Cyprus	0.512	0.350	0.222	6.421	5.571	4.350	0.459	0.580	0.182	0.851	0.101	0.003
Czech Republic	0.439	0.338	0.087	2.675	1.995	0.212	0.933	0.714	0.137	0.679	0.108	0.116
Estonia	0.494	0.371	0.245	2.549	1.995	0.458	0.766	0.606	0.164	0.554	0.111	0.174
Hungary	0.468	0.350	0.152	2.308	1.494	0.540	0.418	0.452	0.084	0.815	0.081	0.157
Latvia	0.527	0.387	0.294	2.330	1.503	0.502	0.337	0.545	0.119	0.827	0.096	0.141
Lithuania	0.638	0.510	0.512	3.109	2.094	0.645	0.378	0.926	0.145	1.016	0.101	0.210
Malta	0.490	0.434	0.358	4.873	4.407	2.168	1.450	0.603	0.186	0.467	0.208	0.012
Poland	0.529	0.378	0.308	3.741	3.179	0.980	1.316	0.767	0.116	0.563	0.115	0.205
Slovakia	0.421	0.181	−0.096	2.801	2.066	0.113	1.561	0.309	0.083	0.735	0.109	0.155
Slovenia	0.642	0.324	0.315	10.410	9.929	5.679	2.913	1.036	0.302	0.481	0.208	0.253
Bulgaria	0.550	0.453	0.222	2.073	1.335	0.443	0.329	0.427	0.137	0.738	0.057	0.106
Romania	0.582	0.484	0.359	3.058	2.407	0.708	1.058	0.494	0.147	0.650	0.065	0.091

(continued)

Table A.9 (continued)

Country	Gross Farm Income (euro)/Total output	Farm Net Value Added (euro)/Total output	Farm Net Income (euro)/Total output	Total assets (euro)/Total output	Total fixed assets (euro)/Total output	Land, permanent crops and quotas (euro)/Total output	Buildings (euro)/Total output	Machinery (euro)/Total output	Breeding livestock (euro)/Total output	Total current assets (euro)/Total output	Non-breeding livestock (euro)/Total output	Stock of agricultural products (euro)/Total output
European Union 1	0.571	0.437	0.294	4.335	3.625	2.414	0.546	0.448	0.217	0.710	0.157	0.116
European Union 2	0.592	0.455	0.295	4.811	3.958	2.668	0.646	0.466	0.178	0.853	0.134	0.121

Country	Other circulating capital (euro)/Total output	Total liabilities (euro)/Total output	Long and medium-term loans (euro)/Total output	Short-term loans (euro)/Total output	Net worth (euro)/Total output	Change in net worth (euro)/Total output	Average farm capital (euro)/Total output	Gross Investment (euro)/Total output	Net Investment (euro)/Total output	Cash Flow (1)(euro)/Total output	Cash Flow (2)(euro)/Total output
Belgium 1	0.087	0.778	0.776	0.002	1.454	0.038	1.528	0.136	0.033	0.387	0.259
Belgium 2	0.113	0.754	0.751	0.003	1.770	0.063	1.532	0.146	0.026	0.392	0.230
Denmark 1	0.324	1.875	1.380	0.495	1.323	0.092	2.499	0.147	0.036	0.189	0.107
Denmark 2	0.554	3.170	2.830	0.341	2.726	0.311	3.421	0.268	0.144	0.122	0.154
France 1	0.459	0.822	0.539	0.283	1.558	0.044	1.909	0.147	0.005	0.381	0.262
France 2	0.530	0.924	0.589	0.335	1.557	0.039	2.102	0.168	-0.007	0.383	0.234
Germany 1	0.414	0.712	0.511	0.201	3.620	0.003	2.064	0.161	-0.001	0.339	0.185
Germany 2	0.382	0.740	0.483	0.256	3.476	0.013	1.724	0.153	0.011	0.312	0.174
Greece 1	0.068	0.088	0.062	0.025	4.001	-0.025	1.849	0.032	-0.088	0.676	0.653
Greece 2	0.096	0.023	0.014	0.009	4.104	-0.132	2.025	0.021	-0.121	0.741	0.717
Ireland 1	0.297	0.369	0.270	0.099	7.076	0.195	2.404	0.122	0.033	0.455	0.352
Ireland 2	0.328	0.475	0.395	0.080	17.462	0.911	3.122	0.135	-0.027	0.630	0.514
Italy 1	0.361	0.084	0.075	0.009	7.287	1.883	2.215	0.068	-0.062	0.553	0.489
Italy 2	0.444	0.076	0.069	0.007	7.029	1.149	2.250	0.066	-0.061	0.569	0.498
Luxembourg 1	0.480	0.708	0.604	0.103	4.624	0.097	2.959	0.308	0.086	0.472	0.201
Luxembourg 2	0.731	1.067	0.921	0.147	5.415	0.080	3.549	0.396	0.108	0.559	0.224
Netherlands 1	0.303	1.145	1.066	0.079	2.127	0.029	1.660	0.184	0.056	0.294	0.156
Netherlands 2	0.411	1.668	1.434	0.233	3.204	0.050	1.630	0.187	0.067	0.225	0.135

(continued)

Table A.9 (continued)

Country	Other circulating capital (euro)/ Total output	Total liabilities (euro) /Total output	Long and medium-term loans (euro) /Total output	Short-term loans (euro)/ Total output	Net worth (euro)/ Total output	Change in net worth (euro)/ Total output	Average farm capital (euro)/ Total output	Gross Investment (euro)/Total output	Net Investment (euro)/Total output	Cash Flow (1)(euro)/ Total output	Cash Flow (2)(euro)/ Total output
Portugal 1	0.262	0.153	0.103	0.050	4.957	0.114	2.615	0.223	0.056	0.364	0.149
Portugal 2	0.324	0.138	0.074	0.064	3.711	0.016	2.256	0.153	-0.014	0.514	0.357
Spain 1	1.364	0.091	0.072	0.019	5.740	0.339	2.973	0.038	-0.057	0.538	0.502
Spain 2	2.017	0.143	0.125	0.016	5.670	0.311	3.044	0.048	-0.012	0.590	0.540
United Kingdom 1	0.297	0.545	0.224	0.321	3.771	0.075	1.533	0.136	0.014	0.319	0.196
United Kingdom 2	0.413	0.653	0.308	0.345	4.786	0.187	1.563	0.159	0.037	0.304	0.164
Austria 1	0.338	0.549	0.388	0.161	5.033	0.012	4.064	0.280	0.049	0.661	0.417
Austria 2	0.847	0.616	0.447	0.169	5.287	0.051	4.609	0.285	0.050	0.636	0.385
Finland 1	0.482	1.012	0.926	0.085	2.790	0.038	2.406	0.275	0.058	0.609	0.327
Finland 2	0.616	1.173	1.097	0.077	3.290	0.158	3.000	0.371	0.086	0.595	0.289
Sweden 1	0.403	1.307	0.965	0.342	2.781	-0.099	3.158	0.217	-0.045	0.292	0.136
Sweden 2	0.543	1.382	1.114	0.268	2.965	0.097	2.630	0.247	0.039	0.292	0.100
Cyprus	0.746	0.144	0.131	0.010	6.278	-0.107	2.518	0.033	-0.129	0.356	0.307
Czech Republic	0.455	0.615	0.371	0.244	2.059	0.048	2.479	0.114	0.013	0.182	0.070
Estonia	0.269	0.709	0.439	0.270	1.841	0.125	1.992	0.291	0.168	0.336	0.154
Hungary	0.577	0.641	0.332	0.309	1.667	0.010	1.845	0.124	0.005	0.248	0.180
Latvia	0.590	0.696	0.484	0.212	1.634	0.151	1.720	0.326	0.187	0.377	0.185
Lithuania	0.704	0.469	0.259	0.210	2.641	0.295	2.283	0.371	0.243	0.581	0.338
Malta	0.247	0.227	0.135	0.092	4.647	0.024	2.772	0.105	0.049	0.410	0.320
Poland	0.243	0.349	0.245	0.104	3.392	0.104	2.819	0.161	0.010	0.429	0.276
Slovakia	0.472	0.231	0.119	0.112	2.570	-0.110	2.786	0.159	-0.080	0.127	-0.012
Slovenia	0.019	0.201	0.196	0.005	10.209	0.122	4.915	0.400	0.082	0.534	0.151
Bulgaria	0.576	0.393	0.255	0.138	1.680	0.188	1.698	0.205	0.108	0.276	0.113

(continued)

Table A.9 (continued)

Country	Other circulating capital (euro)/ Total output	Total liabilities (euro) /Total output	Long and medium-term loans (euro) /Total output	Short-term loans (euro)/ Total output	Net worth (euro)/ Total output	Change in net worth (euro)/ Total output	Average farm capital (euro)/ Total output	Gross Investment (euro)/Total output	Net Investment (euro)/Total output	Cash Flow (1)(euro)/ Total output	Cash Flow (2)(euro)/ Total output
Romania	0.495	0.094	0.056	0.038	2.963	0.245	2.318	0.051	−0.047	0.348	0.304
European Union 1	0.437	0.614	0.446	0.168	3.721	0.332	2.070	0.129	−0.005	0.409	0.299
European Union 2	0.597	0.709	0.526	0.183	4.101	0.254	2.236	0.140	0.002	0.418	0.306

Table A.10 Indicators among subsidies values and the total utilized agricultural area

Country	Balance current subsidies and taxes (euro)/Total utilised agricultural area	Total subsidies—excluding on investments (euro)/Total utilised agricultural area	Environmental subsidies (euro)/Total utilised agricultural area	LFA subsidies (euro)/Total utilised agricultural area	Total support for rural development (euro)/Total utilised agricultural area	Other rural development payments (euro)/Total utilised agricultural area	Other subsidies (euro)/Total utilised agricultural area
Belgium 1	258.806	216.222	0.202	0.000	0.202	0.000	29.640
Belgium 2	433.481	433.105	22.919	8.944	33.055	1.187	17.620
Denmark 1	163.334	198.458	2.651	0.000	2.651	0.000	14.104
Denmark 2	324.147	373.233	12.836	0.433	13.491	0.219	8.171
France 1	159.325	193.025	3.670	0.000	3.670	0.000	12.375
France 2	319.053	343.954	16.465	17.908	34.404	0.030	10.942
Germany 1	228.255	202.923	16.117	0.000	16.117	0.000	37.621
Germany 2	387.389	379.197	35.695	18.304	54.288	0.288	3.253
Greece 1	379.707	355.521	0.059	0.000	0.059	0.000	3.875
Greece 2	724.675	732.308	11.295	37.833	62.847	13.554	8.623
Ireland 1	151.396	153.392	7.822	0.000	7.822	0.000	11.559
Ireland 2	373.071	373.828	60.015	45.947	111.850	5.883	2.897
Italy 1	112.024	142.133	10.220	0.000	10.367	0.000	10.004
Italy 2	339.033	356.382	33.212	8.186	48.701	7.191	17.096
Luxembourg 1	247.411	172.520	25.216	0.000	25.216	0.000	67.609
Luxembourg 2	561.343	491.392	100.202	120.443	220.826	0.182	17.856
Netherlands 1	146.240	64.214	7.111	0.000	7.281	0.000	6.706
Netherlands 2	280.351	366.844	41.030	0.000	43.482	2.452	35.929
Portugal 1	82.557	90.861	9.348	0.000	9.348	0.000	11.342
Portugal 2	190.664	201.591	23.411	26.334	51.874	2.118	1.742

(continued)

Table A.10 (continued)

Country	Balance current subsidies and taxes (euro)/ Total utilised agricultural area	Total subsidies—excluding on investments (euro)/Total utilised agricultural area	Environmental subsidies (euro)/Total utilised agricultural area	LFA subsidies (euro)/Total utilised agricultural area	Total support for rural development (euro)/Total utilised agricultural area	Other rural development payments (euro)/Total utilised agricultural area	Other subsidies (euro)/Total utilised agricultural area
Spain 1	107.284	115.196	0.032	0.000	0.041	0.000	1.216
Spain 2	215.973	215.133	3.545	5.692	10.401	1.063	3.053
United Kingdom 1	147.337	150.957	3.978	0.000	3.996	0.000	3.057
United Kingdom 2	271.621	277.616	26.940	15.477	43.944	1.526	9.747
Austria 1	658.078	630.281	225.941	0.000	225.941	0.000	74.082
Austria 2	633.714	574.981	221.199	78.498	300.946	1.255	3.792
Finland 1	801.270	809.443	136.572	0.000	136.572	0.000	153.747
Finland 2	878.541	882.781	145.835	215.428	366.059	4.789	23.533
Sweden 1	212.597	214.146	38.269	0.000	38.269	0.000	26.122
Sweden 2	333.711	335.091	74.314	18.975	97.425	4.136	4.502
Cyprus	521.014	525.557	4.556	46.200	51.128	0.372	99.207
Czech Republic	242.042	256.282	30.201	25.720	57.396	1.476	44.488
Estonia	137.574	138.540	36.018	10.314	51.755	5.425	18.480
Hungary	216.075	230.330	27.560	0.647	29.952	1.739	20.262
Latvia	172.603	178.282	22.139	25.517	57.666	10.007	19.514
Lithuania	149.758	156.041	0.915	23.511	43.271	18.839	16.559
Malta	3010.177	3032.700	24.136	215.424	675.143	435.635	145.902

(continued)

Table A.10 (continued)

Country	Balance current subsidies and taxes (euro)/Total utilised agricultural area	Total subsidies—excluding on investments (euro)/Total utilised agricultural area	Environmental subsidies (euro)/Total utilised agricultural area	LFA subsidies (euro)/Total utilised agricultural area	Total support for rural development (euro)/Total utilised agricultural area	Other rural development payments (euro)/Total utilised agricultural area	Other subsidies (euro)/Total utilised agricultural area
Poland	194.936	208.996	11.938	20.186	48.886	16.762	62.622
Slovakia	195.291	211.089	29.917	53.160	83.117	0.039	6.721
Slovenia	501.837	557.474	127.509	106.520	269.075	34.999	31.383
Bulgaria	130.353	135.022	0.403	0.106	11.322	10.814	35.817
Romania	142.500	160.085	0.524	0.524	1.129	0.109	51.785
European Union 1	163.874	172.348	8.949	0.000	8.973	0.000	13.243
European Union 2	313.302	322.979	28.175	20.177	51.358	2.977	13.622

Country	Subsidies on intermediate consumption (euro)/Total utilised agricultural area	Subsidies on external factors (euro)/Total utilised agricultural area	Decoupled payments (euro)/Total utilised agricultural area	Single Farm payment (euro)/Total utilised agricultural area	Single Area payment (euro)/Total utilised agricultural area	Additional aid (euro)/Total utilised agricultural area	Support_Art68 (euro)/Total utilised agricultural area
Belgium 1	0.069	77.550	0.000	0.000	0.000	0.000	0.000
Belgium 2	0.017	45.228	150.216	148.640	0.000	1.579	0.000
Denmark 1	5.165	14.171	0.000	0.000	0.000	0.000	0.000
Denmark 2	0.626	8.438	174.240	173.209	0.000	1.028	0.000
France 1	0.000	0.000	0.000	0.000	0.000	0.000	0.000
France 2	0.000	0.000	88.219	87.185	0.000	1.032	0.000

(continued)

Table A.10 (continued)

Country	Subsidies on intermediate consumption (euro)/Total utilised agricultural area	Subsidies on external factors (euro)/Total utilised agricultural area	Decoupled payments (euro)/Total utilised agricultural area	Single Farm payment (euro)/Total utilised agricultural area	Single Area payment (euro)/Total utilised agricultural area	Additional aid (euro)/Total utilised agricultural area	Support_Art68 (euro)/Total utilised agricultural area
Germany 1	13.056	8.835	0.000	0.000	0.000	0.000	0.000
Germany 2	16.975	8.409	160.755	159.852	0.000	0.904	0.000
Greece 1	0.483	0.000	0.000	0.000	0.000	0.000	0.000
Greece 2	0.277	0.016	254.129	246.883	0.000	7.261	0.000
Ireland 1	0.599	0.000	0.000	0.000	0.000	0.000	0.000
Ireland 2	0.748	0.000	148.693	147.165	0.000	1.528	0.000
Italy 1	0.133	1.049	0.000	0.000	0.000	0.000	0.000
Italy 2	0.026	0.257	131.217	129.089	0.000	2.134	0.073
Luxembourg 1	4.764	4.387	0.000	0.000	0.000	0.000	0.000
Luxembourg 2	13.770	2.913	135.080	134.209	0.000	0.871	0.000
Netherlands 1	0.000	4.080	0.000	0.000	0.000	0.000	0.000
Netherlands 2	0.627	1.211	124.529	122.775	0.000	1.754	0.000
Portugal 1	4.698	0.112	0.000	0.000	0.000	0.000	0.000
Portugal 2	1.229	0.122	44.491	43.452	0.000	1.036	0.000
Spain 1	1.854	0.087	0.000	0.000	0.000	0.000	0.000
Spain 2	1.161	0.301	53.243	51.933	0.000	1.310	0.346
United Kingdom 1	0.116	0.026	0.000	0.000	0.000	0.000	0.000

(continued)

Table A.10 (continued)

Country	Subsidies on intermediate consumption (euro)/Total utilised agricultural area	Subsidies on external factors (euro)/Total utilised agricultural area	Decoupled payments (euro)/Total utilised agricultural area	Single Farm payment (euro)/Total utilised agricultural area	Single Area payment (euro)/Total utilised agricultural area	Additional aid (euro)/Total utilised agricultural area	Support_Art68 (euro)/Total utilised agricultural area
United Kingdom 2	0.093	0.009	114.254	113.790	0.000	0.464	0.000
Austria 1	7.007	8.966	0.000	0.000	0.000	0.000	0.000
Austria 2	15.096	10.092	112.049	109.932	0.000	2.117	0.000
Finland 1	0.000	0.000	0.000	0.000	0.000	0.000	0.000
Finland 2	0.000	0.000	89.757	88.244	0.000	1.513	0.000
Sweden 1	0.000	0.932	0.000	0.000	0.000	0.000	0.000
Sweden 2	0.000	4.035	104.960	104.152	0.000	0.809	0.000
Cyprus	0.000	0.000	123.904	0.000	123.904	0.000	0.000
Czech Republic	18.906	5.492	95.634	0.000	95.634	0.000	0.000
Estonia	0.467	0.078	42.670	0.000	42.670	0.000	0.000
Hungary	19.931	9.305	100.293	0.000	100.293	0.000	0.000
Latvia	12.714	3.427	31.510	0.000	31.510	0.000	0.000
Lithuania	0.676	2.004	55.437	0.000	55.437	0.000	0.000
Malta	0.000	0.000	211.004	211.004	0.000	0.000	0.000
Poland	4.776	0.000	77.784	0.000	77.784	0.000	0.000
Slovakia	1.203	0.001	73.712	0.000	73.712	0.000	0.000
Slovenia	9.103	0.000	126.432	126.432	0.000	0.000	0.000
Bulgaria	15.925	1.293	61.258	0.000	61.233	0.025	0.000

(continued)

Table A.10 (continued)

Country	Subsidies on intermediate consumption (euro)/Total utilised agricultural area	Subsidies on external factors (euro)/Total utilised agricultural area	Decoupled payments (euro)/Total utilised agricultural area	Single Farm payment (euro)/Total utilised agricultural area	Single Area payment (euro)/Total utilised agricultural area	Additional aid (euro)/Total utilised agricultural area	Support_Art68 (euro)/Total utilised agricultural area
Romania	15.077	0.000	61.853	0.000	61.853	0.000	0.000
European Union 1	2.453	2.810	0.000	0.000	0.000	0.000	0.000
European Union 2	3.871	2.346	94.511	83.346	10.108	1.056	0.053

Table A.11 Indicators among subsidies values and the total output

Country	Balance current subsidies and taxes (euro)/Total output	Total subsidies—excluding on investments (euro)/Total output	Environmental subsidies (euro)/Total output	LFA subsidies (euro)/Total output	Total support for rural development (euro)/Total output	Other rural development payments (euro)/Total output	Other subsidies (euro)/Total output
Belgium 1	0.065	0.054	0.000	0.000	0.000	0.000	0.007
Belgium 2	0.104	0.103	0.005	0.002	0.008	0.000	0.004
Denmark 1	0.063	0.076	0.001	0.000	0.001	0.000	0.005
Denmark 2	0.103	0.119	0.004	0.000	0.004	0.000	0.003
France 1	0.097	0.117	0.002	0.000	0.002	0.000	0.008
France 2	0.183	0.197	0.009	0.010	0.020	0.000	0.006
Germany 1	0.103	0.093	0.007	0.000	0.007	0.000	0.017
Germany 2	0.174	0.171	0.016	0.008	0.025	0.000	0.002
Greece 1	0.160	0.150	0.000	0.000	0.000	0.000	0.002
Greece 2	0.277	0.280	0.004	0.014	0.024	0.005	0.003
Ireland 1	0.174	0.176	0.010	0.000	0.010	0.000	0.013
Ireland 2	0.423	0.423	0.068	0.052	0.126	0.007	0.003
Italy 1	0.047	0.059	0.004	0.000	0.004	0.000	0.004
Italy 2	0.109	0.115	0.011	0.003	0.016	0.002	0.006
Luxembourg 1	0.140	0.099	0.015	0.000	0.015	0.000	0.039
Luxembourg 2	0.306	0.268	0.055	0.066	0.120	0.000	0.010
Netherlands 1	0.017	0.008	0.001	0.000	0.001	0.000	0.001
Netherlands 2	0.026	0.035	0.004	0.000	0.004	0.000	0.003
Portugal 1	0.100	0.110	0.011	0.000	0.011	0.000	0.014
Portugal 2	0.210	0.222	0.026	0.029	0.057	0.002	0.002

(continued)

Table A.11 (continued)

Country	Balance current subsidies and taxes (euro)/Total output	Total subsidies—excluding on investments (euro)/Total output	Environmental subsidies (euro)/Total output	LFA subsidies (euro)/Total output	Total support for rural development (euro)/Total output	Other rural development payments (euro)/Total output	Other subsidies (euro)/Total output
Spain 1	0.111	0.119	0.000	0.000	0.000	0.000	0.001
Spain 2	0.165	0.165	0.003	0.004	0.008	0.001	0.002
United Kingdom 1	0.127	0.130	0.003	0.000	0.003	0.000	0.003
United Kingdom 2	0.209	0.214	0.020	0.012	0.033	0.001	0.008
Austria 1	0.329	0.315	0.114	0.000	0.114	0.000	0.037
Austria 2	0.333	0.302	0.117	0.041	0.159	0.001	0.002
Finland 1	0.587	0.593	0.100	0.000	0.100	0.000	0.113
Finland 2	0.621	0.624	0.103	0.152	0.258	0.003	0.017
Sweden 1	0.198	0.200	0.038	0.000	0.038	0.000	0.020
Sweden 2	0.248	0.249	0.056	0.014	0.073	0.003	0.003
Cyprus	0.150	0.151	0.001	0.014	0.015	0.000	0.028
Czech Republic	0.207	0.219	0.026	0.022	0.049	0.001	0.036
Estonia	0.239	0.241	0.063	0.018	0.090	0.010	0.031
Hungary	0.187	0.199	0.023	0.001	0.025	0.002	0.016
Latvia	0.287	0.297	0.036	0.043	0.097	0.017	0.031
Lithuania	0.260	0.271	0.001	0.042	0.076	0.032	0.026
Malta	0.174	0.175	0.001	0.012	0.039	0.025	0.009
Poland	0.141	0.151	0.009	0.015	0.035	0.012	0.045

(continued)

Table A.11 (continued)

Country	Balance current subsidies and taxes (euro)/Total output	Total subsidies—excluding on investments (euro)/Total output	Environmental subsidies (euro)/Total output	LFA subsidies (euro)/Total output	Total support for rural development (euro)/Total output	Other rural development payments (euro)/Total output	Other subsidies (euro)/Total output
Slovakia	0.261	0.282	0.040	0.071	0.111	0.000	0.009
Slovenia	0.305	0.340	0.079	0.068	0.164	0.018	0.017
Bulgaria	0.172	0.178	0.001	0.000	0.016	0.015	0.047
Romania	0.133	0.149	0.000	0.000	0.001	0.000	0.048
European Union 1	0.094	0.099	0.005	0.000	0.005	0.000	0.008
European Union 2	0.168	0.174	0.015	0.011	0.028	0.002	0.007

Country	Subsidies on intermediate consumption (euro)/Total output	Subsidies on external factors (euro)/Total output	Decoupled payments (euro)/Total output	Single Farm payment (euro)/Total output	Single Area payment (euro)/Total output	Additional aid (euro)/Total output	Support_Art68 (euro)/Total output
Belgium 1	0.000	0.019	0.000	0.000	0.000	0.000	0.000
Belgium 2	0.000	0.011	0.035	0.034	0.000	0.000	0.000
Denmark 1	0.002	0.005	0.000	0.000	0.000	0.000	0.000
Denmark 2	0.000	0.003	0.050	0.050	0.000	0.000	0.000
France 1	0.000	0.000	0.000	0.000	0.000	0.000	0.000
France 2	0.000	0.000	0.049	0.048	0.000	0.001	0.000
Germany 1	0.006	0.004	0.000	0.000	0.000	0.000	0.000
Germany 2	0.008	0.004	0.070	0.070	0.000	0.000	0.000
Greece 1	0.000	0.000	0.000	0.000	0.000	0.000	0.000

(continued)

Table A.11 (continued)

Country	Subsidies on intermediate consumption (euro)/ Total output	Subsidies on external factors (euro)/Total output	Decoupled payments (euro)/Total output	Single Farm payment (euro)/Total output	Single Area payment (euro)/Total output	Additional aid (euro)/ Total output	Support_Art68 (euro)/Total output
Greece 2	0.000	0.000	0.098	0.095	0.000	0.003	0.000
Ireland 1	0.001	0.000	0.000	0.000	0.000	0.000	0.000
Ireland 2	0.001	0.000	0.165	0.164	0.000	0.002	0.000
Italy 1	0.000	0.000	0.000	0.000	0.000	0.000	0.000
Italy 2	0.000	0.000	0.039	0.038	0.000	0.001	0.000
Luxembourg 1	0.003	0.002	0.000	0.000	0.000	0.000	0.000
Luxembourg 2	0.008	0.002	0.071	0.070	0.000	0.000	0.000
Netherlands 1	0.000	0.000	0.000	0.000	0.000	0.000	0.000
Netherlands 2	0.000	0.000	0.010	0.010	0.000	0.000	0.000
Portugal 1	0.006	0.000	0.000	0.000	0.000	0.000	0.000
Portugal 2	0.001	0.000	0.049	0.048	0.000	0.001	0.000
Spain 1	0.002	0.000	0.000	0.000	0.000	0.000	0.000
Spain 2	0.001	0.000	0.040	0.039	0.000	0.001	0.000
United Kingdom 1	0.000	0.000	0.000	0.000	0.000	0.000	0.000
United Kingdom 2	0.000	0.000	0.085	0.085	0.000	0.000	0.000
Austria 1	0.004	0.005	0.000	0.000	0.000	0.000	0.000
Austria 2	0.008	0.005	0.055	0.054	0.000	0.001	0.000
Finland 1	0.000	0.000	0.000	0.000	0.000	0.000	0.000
Finland 2	0.000	0.000	0.059	0.058	0.000	0.001	0.000
Sweden 1	0.000	0.001	0.000	0.000	0.000	0.000	0.000

(continued)

Table A.11 (continued)

Country	Subsidies on intermediate consumption (euro)/ Total output	Subsidies on external factors (euro)/Total output	Decoupled payments (euro)/Total output	Single Farm payment (euro)/Total output	Single Area payment (euro)/Total output	Additional aid (euro)/ Total output	Support_Art68 (euro)/Total output
Sweden 2	0.000	0.003	0.072	0.072	0.000	0.001	0.000
Cyprus	0.000	0.000	0.035	0.000	0.035	0.000	0.000
Czech Republic	0.016	0.005	0.082	0.000	0.082	0.000	0.000
Estonia	0.001	0.000	0.074	0.000	0.074	0.000	0.000
Hungary	0.017	0.008	0.087	0.000	0.087	0.000	0.000
Latvia	0.021	0.006	0.052	0.000	0.052	0.000	0.000
Lithuania	0.001	0.003	0.095	0.000	0.095	0.000	0.000
Malta	0.000	0.000	0.012	0.012	0.000	0.000	0.000
Poland	0.003	0.000	0.057	0.000	0.057	0.000	0.000
Slovakia	0.002	0.000	0.099	0.000	0.099	0.000	0.000
Slovenia	0.006	0.000	0.063	0.063	0.000	0.000	0.000
Bulgaria	0.021	0.002	0.081	0.000	0.081	0.000	0.000
Romania	0.014	0.000	0.058	0.000	0.058	0.000	0.000
European Union 1	0.001	0.002	0.000	0.000	0.000	0.000	0.000
European Union 2	0.002	0.001	0.051	0.045	0.005	0.001	0.000

Table A.12 Ratios among the animal subsidies and the total subsidies on livestock

Country	Subsidies dairying/Total subsidies on livestock	Subsidies other cattle/Total subsidies on livestock	Subsidies sheep and goats/Total subsidies on livestock	Other livestock subsidies/Total subsidies on livestock
Belgium 1				
Belgium 2	0.046	*0.952*	0.000	0.001
Denmark 1	0.126	0.850	0.023	0.001
Denmark 2	−0.034	*1.023*	0.009	0.002
France 1				
France 2	0.027	0.803	0.084	0.086
Germany 1	0.323	0.474	0.036	0.167
Germany 2	−0.037	0.400	0.020	*0.618*
Greece 1	0.012	0.134	*0.843*	0.010
Greece 2	0.092	0.194	0.683	0.073
Ireland 1	−0.011	0.501	0.351	0.158
Ireland 2	0.284	0.670	0.045	0.000
Italy 1	0.031	0.706	0.167	0.021
Italy 2	0.108	0.320	0.158	0.414
Luxembourg 1				
Luxembourg 2	*1.672*	0.497	0.000	−1.170
Netherlands 1	−0.690	0.871	*0.706*	0.112
Netherlands 2	0.007	0.812	0.068	0.113
Portugal 1	0.106	0.366	0.456	0.070
Portugal 2	0.144	0.633	0.222	0.001
Spain 1	0.012	0.226	*0.740*	0.022
Spain 2	0.025	0.579	0.377	0.019
United Kingdom 1	0.036	0.438	0.515	0.010
United Kingdom 2				
Austria 1	0.000	0.165	0.000	*0.835*
Austria 2	0.034	*0.880*	0.001	0.085
Finland 1	*0.530*	0.196	0.010	0.263
Finland 2	0.506	0.282	0.014	0.198
Sweden 1				
Sweden 2	0.119	0.681	0.005	0.195
Cyprus	0.171	0.158	0.664	0.007
Czech Republic	0.000	0.532	0.015	0.453
Estonia	0.091	0.413	0.131	0.365
Hungary	0.162	0.215	0.217	0.407
Latvia	0.460	0.321	0.056	0.163

(continued)

Table A.12 (continued)

Country	Subsidies dairying/Total subsidies on livestock	Subsidies other cattle/Total subsidies on livestock	Subsidies sheep and goats/Total subsidies on livestock	Other livestock subsidies/Total subsidies on livestock
Lithuania	0.200	0.773	0.020	0.008
Malta	0.165	0.026	0.000	*0.808*
Poland				
Slovakia	0.000	0.233	0.157	0.610
Slovenia	0.091	0.835	0.074	0.001
Bulgaria	*0.686*	0.000	0.300	0.011
Romania	0.214	0.017	0.205	0.563
European Union 1	0.024	0.439	0.437	0.100
European Union 2	0.099	0.646	0.151	0.104

Table A.13 Indicators among labour values and the total utilized agricultural area

Country	Labour input (h)/Total utilised agricultural area	Unpaid labour input (h)/Total utilised agricultural area	Paid labour Input (h)/Total utilised agricultural area	Total labour input (awu)/Total utilised agricultural area	Unpaid labour input (awu)/Total utilised agricultural area	Paid labour input (awu)/Total utilised agricultural area
Belgium 1	158.693	141.690	17.003	0.057	0.050	0.007
Belgium 2	121.217	102.084	19.133	0.045	0.037	0.008
Denmark 1	57.211	40.749	16.462	0.029	0.020	0.009
Denmark 2	40.937	25.151	15.786	0.021	0.012	0.009
France 1	71.441	59.601	11.840	0.031	0.026	0.005
France 2	45.454	33.774	11.680	0.026	0.019	0.007
Germany 1	101.566	81.092	20.474	0.044	0.035	0.009
Germany 2	64.109	41.954	22.155	0.029	0.019	0.010
Greece 1	465.496	425.374	40.122	0.279	0.258	0.020
Greece 2	422.401	370.902	51.500	0.183	0.161	0.023
Ireland 1	75.216	68.210	7.006	0.033	0.030	0.003
Ireland 2	59.016	55.156	3.860	0.027	0.026	0.002
Italy 1	272.248	250.183	22.065	0.117	0.108	0.009
Italy 2	192.171	156.980	35.190	0.086	0.069	0.017
Luxembourg 1	70.260	64.109	6.150	0.030	0.028	0.003
Luxembourg 2	49.804	42.526	7.278	0.023	0.019	0.003
Netherlands 1	204.537	146.325	58.212	0.093	0.061	0.032
Netherlands 2	181.236	110.561	70.675	0.082	0.047	0.035
Portugal 1	297.641	249.480	48.161	0.124	0.104	0.020
Portugal 2	158.895	133.532	25.363	0.072	0.061	0.012
Spain 1	98.069	75.290	22.778	0.047	0.036	0.011

(continued)

Table A.13 (continued)

Country	Labour input (h)/Total utilised agricultural area	Unpaid labour input (h)/Total utilised agricultural area	Paid labour Input (h)/Total utilised agricultural area	Total labour input (awu)/Total utilised agricultural area	Unpaid labour input (awu)/Total utilised agricultural area	Paid labour input (awu)/Total utilised agricultural area
Spain 2	92.222	71.436	20.785	0.045	0.035	0.010
United Kingdom 1	45.666	26.774	18.892	0.019	0.011	0.008
United Kingdom 2	36.590	21.721	14.869	0.015	0.009	0.007
Austria 1	161.741	153.912	7.829	0.073	0.069	0.004
Austria 2	115.957	108.888	7.069	0.051	0.048	0.003
Finland 1	109.227	98.326	10.901	0.052	0.046	0.006
Finland 2	67.818	59.074	8.744	0.033	0.028	0.005
Sweden 1	41.638	37.918	3.720	0.018	0.017	0.002
Sweden 2	34.337	28.599	5.738	0.016	0.013	0.003
Cyprus	386.253	294.039	92.214	0.171	0.130	*0.041*
Czech Republic	70.548	12.447	58.101	0.034	0.006	0.029
Estonia	51.723	25.177	26.547	0.023	0.011	0.012
Hungary	78.285	29.789	48.496	0.036	0.014	0.022
Latvia	78.921	50.273	28.648	0.039	0.024	0.015
Lithuania	85.493	66.964	18.529	0.039	0.030	0.009
Malta	*1466.850*	*1223.763*	*243.087*	*0.555*	*0.448*	*0.107*
Poland	226.711	195.737	30.974	0.103	0.089	0.014
Slovakia	58.878	4.704	54.173	0.031	0.002	0.029
Slovenia	298.343	287.038	11.306	0.157	0.151	0.006

(continued)

Table A.13 (continued)

Country	Labour input (h)/Total utilised agricultural area	Unpaid labour input (h)/Total utilised agricultural area	Paid labour Input (h)/Total utilised agricultural area	Total labour input (awu)/Total utilised agricultural area	Unpaid labour input (awu)/Total utilised agricultural area	Paid labour input (awu)/Total utilised agricultural area
Bulgaria	166.159	83.220	82.938	0.089	0.045	0.045
Romania	393.257	317.752	75.505	0.162	0.129	0.034
European Union 1	124.557	104.513	20.044	0.057	0.048	0.009
European Union 2	103.713	80.552	23.160	0.048	0.037	0.011

Table A.14 Indicators among the labour values and the total output

Country	Labour input (h)/ Total output	Unpaid labour input (h)/Total output	Paid labour Input (h)/ Total output	Total labour input (awu)/ Total output	Unpaid labour input (awu)/ Total output	Paid labour input (awu)/ Total output
Belgium 1	0.039	0.035	0.004	0.000	0.000	0.000
Belgium 2	0.029	0.025	0.005	0.000	0.000	0.000
Denmark 1	0.022	0.015	0.006	0.000	0.000	0.000
Denmark 2	0.013	0.008	0.005	0.000	0.000	0.000
France 1	0.042	0.035	0.007	0.000	0.000	0.000
France 2	0.026	0.019	0.007	0.000	0.000	0.000
Germany 1	0.042	0.034	0.009	0.000	0.000	0.000
Germany 2	0.029	0.019	0.010	0.000	0.000	0.000
Greece 1	0.197	0.180	0.017	*0.000*	*0.000*	0.000
Greece 2	0.161	0.142	0.020	0.000	0.000	0.000
Ireland 1	0.087	0.079	0.008	0.000	0.000	0.000
Ireland 2	0.067	0.063	0.004	0.000	0.000	0.000
Italy 1	0.114	0.105	0.009	0.000	0.000	0.000
Italy 2	0.062	0.051	0.011	0.000	0.000	0.000
Luxembourg 1	0.038	0.035	0.003	0.000	0.000	0.000
Luxembourg 2	0.027	0.023	0.004	0.000	0.000	0.000
Netherlands 1	0.024	0.017	0.007	0.000	0.000	0.000
Netherlands 2	0.018	0.011	0.007	0.000	0.000	0.000
Portugal 1	*0.365*	*0.306*	0.059	*0.000*	*0.000*	0.000
Portugal 2	0.176	0.148	0.028	0.000	0.000	0.000
Spain 1	0.099	0.076	0.023	0.000	0.000	0.000
Spain 2	0.071	0.055	0.016	0.000	0.000	0.000
United Kingdom 1	0.039	0.023	0.016	0.000	0.000	0.000
United Kingdom 2	0.028	0.017	0.011	0.000	0.000	0.000
Austria 1	0.081	0.078	0.004	0.000	0.000	0.000
Austria 2	0.061	0.058	0.004	0.000	0.000	0.000
Finland 1	0.080	0.072	0.008	0.000	0.000	0.000
Finland 2	0.048	0.042	0.006	0.000	0.000	0.000
Sweden 1	0.038	0.035	0.003	0.000	0.000	0.000
Sweden 2	0.026	0.022	0.004	0.000	0.000	0.000
Cyprus	0.111	0.084	0.026	0.000	0.000	0.000
Czech Republic	0.061	0.011	0.050	0.000	0.000	0.000
Estonia	0.091	0.044	0.047	0.000	0.000	0.000

(continued)

Table A.14 (continued)

Country	Labour input (h)/ Total output	Unpaid labour input (h)/Total output	Paid labour Input (h)/ Total output	Total labour input (awu)/ Total output	Unpaid labour input (awu)/ Total output	Paid labour input (awu)/ Total output
Hungary	0.068	0.026	0.042	0.000	0.000	0.000
Latvia	0.135	0.086	0.049	0.000	0.000	0.000
Lithuania	0.152	0.119	0.033	0.000	0.000	0.000
Malta	0.085	0.071	0.014	0.000	0.000	0.000
Poland	0.165	0.142	0.022	0.000	0.000	0.000
Slovakia	0.080	0.006	*0.074*	0.000	0.000	*0.000*
Slovenia	0.193	*0.185*	0.007	0.000	0.000	0.000
Bulgaria	*0.218*	0.109	*0.109*	0.000	0.000	*0.000*
Romania	*0.365*	*0.295*	*0.070*	*0.000*	*0.000*	*0.000*
European Union 1	0.070	0.058	0.011	0.000	0.000	0.000
European Union 2	0.056	0.043	0.012	0.000	0.000	0.000

Table A.15 Several indicators among outputs, inputs and subsidies

Country	Total output/Total utilised agricultural area	Total Inputs/Total output	Total Inputs/Total utilised agricultural area	Total output crops and crop production/Total output	Total output livestock and livestock products/Total output	Total subsidies on crops/Total utilised agricultural area	Total subsidies on livestock/Total livestock units	Total subsidies on crops/Total output crops and crop production	Total subsidies on livestock/Total output livestock and livestock products
Belgium 1	4022.587	0.752	3020.641	0.334	0.653	56.170	17.912	0.041	0.021
Belgium 2	4169.935	0.818	3414.585	0.396	0.587	58.513	48.261	0.038	0.053
Denmark 1	2639.130	0.989	2604.994	0.331	0.637	149.949	7.197	0.181	0.007
Denmark 2	3170.051	1.092	3478.833	0.301	0.633	135.235	19.633	0.164	0.017
France 1	1680.747	0.839	1407.634	0.523	0.437	118.281	39.073	0.139	0.046
France 2	1746.839	0.969	1692.558	0.535	0.420	143.388	78.107	0.157	0.092
Germany 1	2379.915	0.911	2154.135	0.340	0.555	114.924	10.398	0.151	0.010
Germany 2	2229.571	0.991	2207.085	0.386	0.520	110.553	22.570	0.141	0.022
Greece 1	2367.723	0.547	1295.164	0.800	0.196	280.703	121.917	0.149	0.152
Greece 2	2627.968	0.646	1694.701	0.755	0.227	352.028	85.636	0.174	0.098
Ireland 1	870.289	0.816	709.470	0.109	0.867	14.856	90.024	0.172	0.157
Ireland 2	884.844	0.973	861.331	0.125	0.847	14.640	76.905	0.146	0.129
Italy 1	2395.263	0.604	1445.905	0.667	0.322	113.228	11.796	0.072	0.010
Italy 2	3142.499	0.655	2056.583	0.664	0.309	138.815	24.713	0.073	0.021
Luxembourg 1	1836.363	0.865	1578.456	0.231	0.710	41.748	20.601	0.103	0.024
Luxembourg 2	1843.814	1.039	1914.233	0.255	0.635	44.182	43.480	0.111	0.049
Netherlands 1	8417.106	0.837	7047.155	0.441	0.530	38.276	1.947	0.010	0.002
Netherlands 2	10381.596	0.904	9436.387	0.501	0.427	65.679	27.119	0.014	0.023
Portugal 1	818.160	0.845	689.043	0.587	0.364	37.455	60.251	0.077	0.097
Portugal 2	909.681	0.851	774.420	0.553	0.392	50.913	102.384	0.100	0.143

(continued)

Table A.15 (continued)

Country	Total output/Total utilised agricultural area	Total Inputs/Total output	Total Inputs/Total utilised agricultural area	Total output crops and crop production/Total output	Total output livestock and livestock products/Total output	Total subsidies on crops/Total utilised agricultural area	Total subsidies on livestock/Total livestock units	Total subsidies on crops/Total output crops and crop production	Total subsidies on livestock/Total output livestock and livestock products
Spain 1	989.558	0.656	650.259	0.635	0.356	86.525	57.501	0.139	0.075
Spain 2	1313.648	0.628	823.029	0.620	0.366	105.861	69.264	0.133	0.085
United Kingdom 1	1170.877	0.928	1087.379	0.396	0.541	65.890	84.825	0.148	0.124
United Kingdom 2	1300.393	1.026	1331.037	0.435	0.512	59.997	57.541	0.112	0.077
Austria 1	1982.943	0.799	1583.310	0.284	0.513	177.729	132.886	0.314	0.130
Austria 2	1912.901	0.871	1664.662	0.285	0.513	72.856	70.415	0.143	0.063
Finland 1	1370.438	1.183	1620.933	0.376	0.591	205.914	431.034	0.409	0.388
Finland 2	1421.427	1.302	1850.447	0.378	0.560	163.555	398.489	0.325	0.304
Sweden 1	1088.921	1.175	1276.453	0.293	0.624	117.759	56.427	0.372	0.051
Sweden 2	1350.560	1.158	1555.524	0.358	0.516	80.452	69.809	0.207	0.064
Cyprus	3512.441	0.938	3290.917	0.510	0.489	119.956	108.571	0.071	0.078
Czech Republic	1162.605	1.128	1310.780	0.533	0.407	23.054	21.907	0.041	0.024
Estonia	573.208	1.048	600.324	0.435	0.487	15.180	29.794	0.062	0.036
Hungary	1167.645	1.048	1217.112	0.568	0.331	37.720	32.176	0.061	0.036
Latvia	597.154	1.045	625.699	0.461	0.469	29.809	67.610	0.111	0.086
Lithuania	581.196	0.853	493.612	0.592	0.391	27.052	35.357	0.085	0.053
Malta	17359.233	0.819	14221.762	0.431	0.561	575.909	155.867	0.078	0.148

(continued)

Table A.15 (continued)

Country	Total output/Total utilised agricultural area	Total Inputs/Total output	Total Inputs/Total utilised agricultural area	Total output crops and crop production/Total output	Total output livestock and livestock products/Total output	Total subsidies on crops/Total utilised agricultural area	Total subsidies on livestock/Total livestock units	Total subsidies on crops/Total output crops and crop production	Total subsidies on livestock/Total output livestock and livestock products
Poland	1389.466	0.825	1143.530	0.516	0.470	14.917	0.013	0.022	0.000
Slovakia	749.669	1.376	1023.702	0.500	0.334	27.986	59.459	0.077	0.075
Slovenia	1652.221	1.024	1688.403	0.401	0.481	32.344	82.663	0.079	0.120
Bulgaria	767.305	0.954	730.170	0.624	0.295	4.626	14.981	0.010	0.021
Romania	1075.938	0.776	835.111	0.600	0.375	3.221	51.624	0.005	0.066
European Union 1	1785.564	0.801	1427.352	0.490	0.467	98.263	43.438	0.116	0.049
European Union 2	1862.804	0.875	1628.801	0.511	0.437	107.869	55.626	0.116	0.060

Table A.16 Crop productions by ha

Country	Cereals (euro/ha)	Vegetables and flowers (euro/ha)	Vineyards (euro/ha)	Olive groves (euro/ha)	Forage crops (euro/ha)
Belgium 1	*1038.906*	22165.975			2.212
Belgium 2	988.949	23838.214			24.979
Denmark 1	794.591	30962.564			34.078
Denmark 2	750.372	42246.831			380.179
France 1	921.775	10848.048	7618.254		10.018
France 2	824.884	14980.196	9768.498	*2413.333*	0.310
Germany 1	848.408	28114.816	11740.013		26.362
Germany 2	766.354	26481.036	*12273.824*		86.438
Greece 1	711.359	11254.683	3681.358	1977.377	*1084.058*
Greece 2	696.429	18901.439	4780.066	2072.064	*1113.751*
Ireland 1	829.301	7553.030			7.973
Ireland 2	852.644				28.678
Italy 1	954.132	13003.168	5106.878	2030.131	508.606
Italy 2	914.221	20935.786	7483.402	*2631.972*	526.840
Luxembourg 1	676.466		*21664.934*		2.368
Luxembourg 2	676.728		*20101.223*		87.297
Netherlands 1	*1085.819*	52477.219			38.514
Netherlands 2	*1003.160*	58039.862			56.853
Portugal 1	540.646	6277.496	1918.319	391.027	148.925
Portugal 2	428.454	8261.485	2553.644	428.611	142.829
Spain 1	421.448	11139.548	1269.521	1062.393	176.495
Spain 2	486.299	17533.006	1789.669	1455.868	142.648
United Kingdom 1	945.407	13331.034			11.480
United Kingdom 2	914.739	15251.358			59.791
Austria 1	675.203	8540.152	5956.571		14.134
Austria 2	623.794	7871.023	6826.921		11.348
Finland 1	425.067	24249.482			19.969
Finland 2	420.713	*59000.711*			23.409
Sweden 1	500.884	3019.299			21.261
Sweden 2	548.981	10776.730			217.735
Cyprus	273.797	18192.075	3264.510	*2074.599*	687.813
Czech Republic	592.927	6642.906	3790.290		225.078
Estonia	311.844	18306.654			120.677
Hungary	578.386	5518.543	4882.819		206.957
Latvia	347.530	4612.159			107.886
Lithuania	384.683	4996.020			164.065

(continued)

Table A.16 (continued)

Country	Cereals (euro/ha)	Vegetables and flowers (euro/ha)	Vineyards (euro/ha)	Olive groves (euro/ha)	Forage crops (euro/ha)
Malta		19438.818	7678.182	1279.167	*896.624*
Poland	465.062	8811.511			54.168
Slovakia	484.080	2476.769	1233.157		139.089
Slovenia	691.964	9126.667	6432.719		320.104
Bulgaria	384.540	12276.931	1036.394		153.850
Romania	452.049	9733.791	3031.111		414.312
European Union 1	791.752	16534.781	4861.528	1463.392	84.923
European Union 2	672.683	20362.576	5950.837	1766.507	114.807

Table A.17 Livestock productions by LU

Country	Cows' milk and milk products (euro/LU)	Beef and veal (euro/LU)	Sheep and goats (euro/LU)	Ewes' and goats' milk (euro/LU)	Pigmeat (euro/LU)	Poultrymeat (euro/LU)	Eggs (euro/LU)
Belgium 1	1472.264	712.336	339.574	222.691	759.017	1022.478	228.772
Belgium 2	1809.396	627.043	356.714	983.384	762.311	1241.183	133.595
Denmark 1	2162.809	564.167	394.922	0.000	787.408	488.274	307.391
Denmark 2	2565.611	494.573	407.001	0.000	878.483	601.874	325.066
France 1	1660.606	551.425	412.252	400.330	790.594	394.516	92.734
France 2	1967.175	526.190	485.815	544.711	743.665	342.286	72.750
Germany 1	1713.962	607.802	318.316	0.718	803.047	462.380	631.932
Germany 2	2140.444	516.584	348.448	24.445	814.359	689.888	389.660
Greece 1	1266.723	608.305	275.828	441.903	1185.919	836.960	593.382
Greece 2	1710.348	474.417	307.881	567.338	934.208	464.659	584.777
Ireland 1	1265.028	399.886	283.616	0.041	578.309	625.206	125.557
Ireland 2	1468.184	400.841	393.891	0.000			
Italy 1	2111.574	773.651	385.596	650.244	997.637	307.214	166.636
Italy 2	2442.530	730.397	395.155	653.430	951.331	154.722	292.982
Luxembourg 1	1837.440	461.486			826.304		
Luxembourg 2	2228.274	424.973	171.675	1069.552	694.708	97.111	2295.773
Netherlands 1	2277.359	559.547	438.098	63.058	852.157	327.374	532.631
Netherlands 2	2469.595	404.857	436.897	1059.184	861.732	384.754	629.994
Portugal 1	1273.660	436.281	224.355	117.741	677.030	541.750	179.414
Portugal 2	1824.431	365.921	254.114	169.680	844.305	380.382	108.487
Spain 1	1323.085	780.313	463.196	250.138	683.291	497.206	145.967

(continued)

Table A.17 (continued)

Country	Cows' milk and milk products (euro/LU)	Beef and veal (euro/LU)	Sheep and goats (euro/LU)	Ewes' and goats' milk (euro/LU)	Pigmeat (euro/LU)	Poultrymeat (euro/LU)	Eggs (euro/LU)
Spain 2	1971.767	679.005	490.468	472.299	559.953	366.937	132.417
United Kingdom 1	1651.560	429.490	309.769	0.000	701.392	248.891	578.661
United Kingdom 2	1920.234	441.337	390.042	1.734	660.146	450.375	477.604
Austria 1	1497.181	685.657	461.877	0.000	861.584	513.154	992.981
Austria 2	1877.390	699.658	591.825	428.591	948.741	814.277	860.874
Finland 1	2331.894	516.761	313.296	10.119	641.105	256.777	494.620
Finland 2	2942.893	537.691	428.249	10.730	772.240	734.469	348.876
Sweden 1	2637.802	399.414	283.856	0.000	827.118	86.974	716.474
Sweden 2	2603.636	443.400	438.135	4.447	815.938	37.654	667.994
Cyprus	2680.273	607.711	864.071	870.111	940.275	762.907	1183.826
Czech Republic	1753.234	404.858	389.268	285.432	780.575	675.357	322.788
Estonia	1578.288	270.873	305.827	19.369	697.546	95.625	1166.524
Hungary	1739.587	432.438	373.245	15.610	738.416	1104.917	224.894
Latvia	1122.314	267.477	381.108	139.514	622.023	214.244	834.939
Lithuania	1035.580	354.532	426.649	18.330	712.557	106.752	699.005
Malta	2024.307	521.856	202.092	2956.805	656.051	579.995	652.514
Poland	1063.866	486.487	271.191	105.368	673.840	1528.698	303.010
Slovakia	1488.622	281.174	136.430	191.144	623.491	519.588	411.159
Slovenia	1378.711	426.566	411.323	31.279	439.359	29.102	316.282
Bulgaria	955.699	319.533	413.556	233.384	664.362	393.670	284.188

(continued)

Table A.17 (continued)

Country	Cows' milk and milk products (euro/LU)	Beef and veal (euro/LU)	Sheep and goats (euro/LU)	Ewes' and goats' milk (euro/LU)	Pigmeat (euro/LU)	Poultrymeat (euro/LU)	Eggs (euro/LU)
Romania	1104.100	366.974	283.992	286.593	657.379	430.756	406.130
European Union 1	1738.918	553.541	347.707	207.985	787.386	381.279	263.993
European Union 2	1978.880	514.933	402.385	332.435	768.840	454.674	263.521

Table A.18 Some productions by ha and by cow

Country	Yield of wheat (100kg/ha)	Yield of maize (100kg/ha)	Stocking density (LU/ha)	Milk yield (Kg/cow)
Belgium 1	75.883	102.047	2.620	5040.540
Belgium 2	81.444	115.569	2.390	6073.276
Denmark 1	70.983		2.663	6352.387
Denmark 2	72.162	67.930	2.213	7817.626
France 1	68.570	82.148	1.334	5532.226
France 2	70.712	88.786	1.293	6290.549
Germany 1	67.067	72.011	1.920	5500.845
Germany 2	70.159	87.351	1.699	6897.613
Greece 1	28.506	107.009	*4.676*	3770.543
Greece 2	29.888	*115.708*	*4.845*	4605.129
Ireland 1	79.132		1.306	4540.633
Ireland 2	*89.614*		1.278	5209.160
Italy 1	48.874	97.686	1.323	5128.141
Italy 2	53.864	101.969	1.489	6099.958
Luxembourg 1	55.458	70.733	1.772	5618.841
Luxembourg 2	59.956	75.478	1.561	6964.471
Netherlands 1	*82.515*	82.291	2.680	6969.443
Netherlands 2	*82.696*	98.895	2.248	7574.522
Portugal 1	17.224	49.213	0.612	4573.954
Portugal 2	16.163	60.449	0.631	6233.402
Spain 1	26.807	89.190	0.865	4601.008
Spain 2	31.354	105.645	0.876	6134.331
United Kingdom 1	75.855		1.127	5847.051
United Kingdom 2	81.415	73.386	1.090	6858.260
Austria 1	52.472	88.422	1.174	4971.020
Austria 2	50.421	100.661	0.918	6015.244
Finland 1	36.556	*192.855*	1.346	6944.200
Finland 2	38.862	38.238	1.252	*8166.878*
Sweden 1	58.140		0.930	*7925.286*
Sweden 2	59.388	109.765	0.992	*8036.393*
Cyprus	30.835	*198.980*	2.608	6640.358
Czech Republic	52.965	72.905	0.873	6334.550
Estonia	29.847		0.480	6460.568
Hungary	44.418	68.802	0.945	6402.247
Latvia	37.072		0.437	4989.653
Lithuania	43.427		0.567	4984.337
Malta			*12.777*	5366.165

(continued)

Table A.18 (continued)

Country	Yield of wheat (100kg/ha)	Yield of maize (100kg/ha)	Stocking density (LU/ha)	Milk yield (Kg/cow)
Poland	50.118	67.123	1.518	4544.545
Slovakia	42.365	64.330	0.503	5727.260
Slovenia	44.505	85.723	1.045	5030.490
Bulgaria	30.017	32.400	1.053	3638.147
Romania	29.387	38.957	1.230	3652.927
European Union 1	62.945	86.196	1.355	5537.638
European Union 2	61.963	86.715	1.273	6348.882

Table A.19 Crop and livestock productions and specific costs by ha and LU, respectively

Country	Total crops output (euro/ha)	Total livestock output/ (euro/LU)	Specific crop costs (euro/ha)	Specific livestock costs (euro/LU)
Belgium 1	1335.182	900.091	435.091	409.000
Belgium 2	1623.700	923.100	521.600	450.200
Denmark 1	914.455	982.636	288.545	482.455
Denmark 2	1023.400	1103.500	306.400	644.200
France 1	913.091	846.909	277.364	291.909
France 2	977.300	854.800	295.700	297.700
Germany 1	842.455	1004.364	283.818	409.273
Germany 2	905.400	1072.600	320.700	456.000
Greece 1	1983.091	792.818	277.727	413.455
Greece 2	2073.400	897.000	369.100	440.800
Ireland 1	94.545	581.727	103.455	178.364
Ireland 2	112.000	612.900	116.100	221.900
Italy 1	1662.091	1213.727	279.727	632.364
Italy 2	2174.800	1176.100	410.400	597.800
Luxembourg 1	426.364	918.455	200.273	284.545
Luxembourg 2	481.500	924.200	223.100	358.700
Netherlands 1	*3756.182*	1131.818	*1149.000*	477.636
Netherlands 2	*5327.000*	*1214.500*	*1577.000*	545.900
Portugal 1	701.091	631.636	110.636	397.273
Portugal 2	720.500	714.000	134.600	426.800
Spain 1	710.364	771.364	128.273	422.455
Spain 2	918.200	809.500	158.000	421.900
United Kingdom 1	477.455	687.182	198.636	297.455
United Kingdom 2	597.500	782.200	246.300	389.800
Austria 1	589.400	1011.200	159.200	355.200
Austria 2	565.400	1150.000	154.600	417.700
Finland 1	564.800	1115.000	245.400	517.600
Finland 2	598.400	*1323.500*	254.400	626.300
Sweden 1	344.400	1140.600	159.200	500.200
Sweden 2	533.400	1088.500	193.300	638.700
Cyprus	2001.500	*1420.000*	429.667	*970.500*
Czech Republic	622.667	928.167	215.333	623.167
Estonia	261.500	843.833	94.333	568.333
Hungary	676.167	968.667	221.167	*692.500*
Latvia	310.833	791.333	93.333	536.833
Lithuania	370.000	715.500	119.000	406.833

(continued)

Table A.19 (continued)

Country	Total crops output (euro/ha)	Total livestock output/ (euro/LU)	Specific crop costs (euro/ha)	Specific livestock costs (euro/LU)
Malta	*7649.500*	1057.167	*1619.333*	*787.500*
Poland	705.333	870.333	229.667	481.667
Slovakia	389.833	765.833	166.333	634.500
Slovenia	696.167	705.167	167.167	458.000
Bulgaria	491.000	686.667	144.000	448.000
Romania	653.333	743.667	166.333	419.333
European Union 1	928.000	882.091	243.727	376.273
European Union 2	1009.000	931.900	274.100	430.600

Table A.20 Average values of the technical efficiency (total specific costs/total output) for the twenty seven former European Union countries

Country	Total specific costs/Total output
Belgium 1	0.408
Belgium 2	0.415
Denmark 1	0.424
Denmark 2	0.454
France 1	0.315
France 2	0.315
Germany 1	0.347
Germany 2	0.364
Greece 1	0.218
Greece 2	0.249
Ireland 1	0.386
Ireland 2	0.436
Italy 1	0.284
Italy 2	0.287
Luxembourg 1	0.330
Luxembourg 2	0.362
Netherlands 1	0.364
Netherlands 2	0.344
Portugal 1	0.364
Portugal 2	0.384
Spain 1	0.325
Spain 2	0.311
United Kingdom 1	0.403
United Kingdom 2	0.444
Austria 1	0.260
Austria 2	0.267
Finland 1	0.454
Finland 2	0.444
Sweden 1	0.419
Sweden 2	0.444
Cyprus	0.454
Czech Republic	0.458
Estonia	*0.493*
Hungary	0.426
Lithuania	0.430
Latvia	0.472
Malta	*0.515*
Poland	0.429
Slovakia	*0.494*

(continued)

Table A.20 (continued)

Country	Total specific costs/Total output
Slovenia	0.388
Bulgaria	0.381
Romania	0.364
European Union 1	0.336
European Union 2	0.349

Printed in the United States
By Bookmasters